THE ATMOSPHERE:
Endangered and Endangering

R. S. Harwood

THE ATMOSPHERE: Endangered and Endangering

William W. Kellogg, Ph.D.

and

Margaret Mead, Ph.D.

Scientific Editors

CASTLE HOUSE PUBLICATIONS LTD.

First published in the UK, 1980

Castle House Publications Ltd
Castle House
27 London Road
Tunbridge Wells TN1 1BX
Kent
England

ISBN 0 7194 0057 0

Printed and bound in Great Britain by
William Clowes (Beccles) Limited, Beccles and London

Organization of this Report

The wide range of subjects dealt with by the Conferees posed problems for the organizers of this report. In order for the busy executive or legislator to identify quickly the main points, we have included a brief *Summary and Recommendations* section. This was prepared by the editors with the help of the rapporteurs and some session chairpersons.

In the body of the report we have set the most significant paragraphs in bold type so that they can be identified easily.

Each appendix is a self-contained contribution dealing with specific points in more detail than in the text. They are included because we believe that they are valuable contributions to the discussion and should be preserved. In the appendixes we also have used bold type to emphasize significant points.

The repetition of some data is not only inevitable, due to the multiple contributors and rapporteurs, but desirable in that it focuses attention on the main subjects from several different points of view. Where possible we have cross-referenced subjects to improve the cohesiveness of our coverage.

Most of the participants did not review this report in its entirety, although individual authors and rapporteurs reviewed their own sections; thus it is possible that some participants will disagree with certain points. However, we worked from taped transcripts and written contributions, and we believe that we have captured something very close to a consensus. Moreover, we believe that the message conveyed on this report enjoys enthusiastic unanimity.

THE EDITORS

Organizing Committee

Harry Davis Bruner
David Garvin
James F. Haggerty
William W. Kellogg
J. F. Lovelock
Margaret Mead
James Peterson
David P. Rall
Walter O. Roberts

Contributors

Mm Jacqueline Beaujeu-Garnier
Centre de Recherches d'Analyse de l'Espace, Paris

Mr Anthony J Broderick
D.O.T. Transportation Systems Centre, Massachusetts

Professor Wallace S Broecker
Lamont-Doherty Geological Observatory of Columbia University

Professor Cyril Brosset
Swedish Water and Air Pollution Research Laboratory

Professor Harrison S Brown
California Institute of Technology

Dr Harry Davis Bruner
Division of Biomedical and Environmental Research, ERDA

Dr David Garvin
National Bureau of Standards, Washington DC

Dr Richard S Greeley
Energy, Resources and Environmental Systems, The MITRE Corporation

Dr Eric Hannah
Department of Physics, Princeton University

Professor John P Holdren
Energy and Resources Program, University of California

Professor John Imbri
Department of Geological Sciences, Brown University

Dr William W Kellogg
National Center for Atmospheric Research, Colorado

Dr Joseph Knox
Physics Department, Lawrence Livermore Laboratory

Dr James E Lovelock

Dr James D McQuigg, Director
Center for Climatic and Environmental Assessment, EDS/NOAA

Dr Margaret Mead

Dr Murray Mitchell Jr
Environmental Data Service, NOAA

Mrs Brooke D Mordy
The Center for the Study of Democratic Institutions, California

Dr Wendell A Mordy
The Center for the Study of Democratic Institutions, California

Dr Toshiichi Okita
Institute of Public Health, Japan

Dr James T Peterson
Meteorology Laboratories, National Environmental Research Center, EPA

Mr Norman W Pirie
Rothamsted Experimental Station, England

Dr David P Rall
National Institute of Environmental Health Sciences, North Carolina

Professor Jack Richardson
National University of Canberra

Dr Walter O Roberts
Aspen Institute for Humanistic Studies, Colorado

Dr G D Robinson
Center for the Environment and Man Inc, Connecticut

Dr Stephen H Schneider
National Center for Atmospheric Research, Colorado

Professor Raymond S Sleeper
University of Tennessee

Mr John Stroud
Simon Frasier University, British Columbia

Dr J Dana Thompson
JAYCOR

Dr Edith Brown Weiss
Department of Civil Engineering and Politics, Princeton University

Ms Barbara West
National Academy of Sciences, Washington DC

Dr George M Woodwell
Marine Biological Laboratory, Massachusetts

Dr Charles A Zraket
The MITRE Corporation

Contents

Foreword

When Dr. Margaret Mead was a Visiting Scholar at the Fogarty International Center, one of her interests focused on the interactions between the world society and its planetary environment. She saw a conflict developing, and yet there was surprisingly little public awareness of the growing problems and few efforts to develop long-term national and international solutions to these problems. She therefore persuaded the Fogarty International Center to sponsor a conference on the atmospheric environment which would explore the ways to maintain it as a healthy place in which to live.

An organizing committee planned the Conference, and its members are listed in these Proceedings. We were fortunate in being able to enlist the help of Dr. William W. Kellogg, of the National Center for Atmospheric Research, to work with Dr. Mead as co-organizer and co-editor of the Proceedings; he is known internationally for his work on climate change and mankind's influence on climate. Four able and dedicated rapporteurs were also enlisted, and this report owes its existence largely to their efforts. They are Mr. Anthony Broderick, Doctors Richard S. Greeley and J. Dana Thompson, and Ms. Barbara West.

It is customary in a foreword to thank the participants, but in this instance we must go beyond the usual expressions of appreciation and say that the success of the Conference stems in a very real sense from the lively interest, support, and intellectual input of all those who attended, especially those who came prepared to make the presentations that are summarized here. It was a hard-working group, deeply concerned with the subject at hand.

The meeting was held at the National Institute of Environmental Health Sciences, Research Triangle Park, North Carolina, at the invitation of its Director, Dr. David P. Rall. We greatly appreciate the care taken with the arrangements and the warm personal hospitality of the hosts.

The Proceedings will be distributed to a wide audience, including many individuals who are in a position to read its message and set policies that will enable society to live in closer harmony with its atmospheric environment. Dr. Mead's Preface (written before the Conference as a "Position Paper") states the purpose of the Conference very forcefully, and makes it clear that many important changes in our way of conducting business, politics, and international relations will have to be made before we can be satisfied that we are behaving responsibly where the health of the atmospheric environment is concerned. Furthermore, scientists have a great deal more homework to do in order to advise policy makers adequately, and many of the more pressing scientific questions are identified here. Some of these scientific questions will require a long time to answer, and some decisions will have to be made before our understanding is complete.

For that reason, we should look upon this Conference as part of a continuing dialogue between scientists and policy makers, a dialogue that must eventually involve the entire world.

MILO D. LEAVITT, JR., M.D.
Director, Fogarty International Center

Preface[1]

SOCIETY AND THE ATMOSPHERIC ENVIRONMENT

We are facing a period when society must make decisions on a planetary scale. Tremendous tankers traverse the seas, supersonic transport aircraft traverse the skies, and nuclear explosions reverberate around the world (both physically and in our consciousness). Whereas in the recent past a whole continent could have been submerged, decimated by plague, or ravaged by earthquakes and the rest of the world remain untouched and unnoticing, today's natural catastrophes and environmental interventions affect the whole of human society—interconnected as it is in reality though not yet politically capable of acting in concert.

As such, manmade interventions depend upon the application of science to technology; scientists become doubly responsible, both for the immediate uses made of their discoveries and for the well-being of their fellow citizens. Whether they be citizens of a free enterprise state, a socialist state, a dictatorship, or a hereditary monarchy, they need information to make decisions, either for an intelligent choice among alternatives or for guidance in carrying out decrees by their ruling group. Even in the most arbitrary and authoritarian forms of government, a comprehension on the part of the leadership and an understanding on the part of the people are both essential. Unless the peoples of the world can begin to understand the immense and long-term consequences of what appear to be small immediate choices—to drill a well, open a road, build a large airplane, make a nuclear test, install a liquid fast breeder reactor, release chemicals which diffuse throughout the atmosphere, or discharge waste in concentrated amounts into the sea—the whole planet may become endangered.

The decisionmakers of the world are beginning to understand many of these issues, but they are trapped in immediacy—how to relieve immediate hunger, how to gain a quick agricultural yield, how to acquire oil at the cheapest rate, how to dispose of waste in some manner that does not entail rebuilding a city or an expense that the existing tax structure will not bear, how to prohibit some form of manufacture that is proving hazardous without creating more depression in already depressed industries. Never before have the governing bodies of the world been faced with decisions so far reaching in their immediate

[1] Prefaces are conventionally written last, as an introduction to the written word. In this case, however, the Preface was written before the Conference and, together with the material by William W. Kellogg in Part 6, served as a position paper to keynote the theme and purpose of the Conference.

consequences and so potentially disastrous and momentous in their long-term consequences. It is inevitable that there will be a clash between those concerned with immediate problems and those who concern themselves with long-term consequences, such as the next 15 years (perhaps) for the victims of certain kinds of chemical industrial processes, the next 25 years for genetic damage to show up in the offspring of today's school children, the next 25 years for the widespread loss of agricultural lands in the tropics, the next 25 to 50 years for possible climatic change, and the next 100 years for the strangulation of our human settlements which have been permitted to aggregate without reference to the condition of the soil, the water, and the air.

For many thousands of years there have been those who have thought in terms of preserving arable and resource land, shorelines, and streams; and ever since human beings began to live in settled communities and to cultivate the land the question of frontiers and the ownership of streams and offshore waters has been a matter for political decisions—and of the willingness of men to defend their land for the safety and future of their families, and of aggressors to attempt to add to what they possessed by inroads on the possessions of others.

The question of the freedom of the seas came much later, and society now faces new relationships between attempts to dominate the sea lanes and attempts to preserve the open oceans as a heritage of all mankind. The Law of the Sea has been essentially the law of the strongest, with a detente established only when there were many nations who benefited by international regulations concerning ports and the prevention of interference with ships at sea.

One has, however, only to follow the discussions of the last few years in the United Nations conferences on the Law of the Sea to see how long-term and short-term political aims conflict, how an interest in the right of navies to sail through straits conflicts with a nation's attempts to protect and exploit its offshore resources of irreplaceable physical resources—oil and minerals—and its replaceable but dangerously depletable biological resources—fish. Rapid about-faces in national policies, conflicts between different economic interests, and the lack of scientific knowledge about the resources of the oceans or the way in which toxic substances are distributed have all played their part in the relative impotence of the Caracas, Vienna, and New York conferences on the Law of the Sea.

At this Conference we are proposing that, before there is a corresponding attempt to develop a "law of the air," the scientific community advise the United Nations (and individual, powerful nation states or aggregations of weaker states) and attempt to arrive at some overview of what is presently known about hazards to the atmosphere from manmade interventions, and how scientific knowledge coupled with intelligent social action can protect the peoples of the world from dangerous and preventable interference with the atmosphere upon which all life depends. We also need warnings about what is *not* known, and about the probability of bad harvests which will alert the nations to build food banks, and about the probability of earthquakes, volcanic eruptions, typhoons, and floods which will be sufficient to persuade them to stop building whole cities on sites where earthquakes and other natural disasters are predictable—even if there is a long and uncertain timespan between such events.

Human beings are endowed individually with an extraordinary ability to forget pain, and this capacity has made human continuance tolerable. When this individual capacity was extended to the community level, so that whole village populations returned to their former sites after a devastation, it may also have been beneficial if they had no other choice than to survive by extreme attachment to a locale and continuation of their traditional way of life. Today, however, this refusal to face the reality of natural disasters and the insistence on denying the probability of their reoccurrence have become definite hindrances to survival.

What we need from scientists are estimates, presented with sufficient conservatism and plausibility but at the same time as free as possible from internal disagreements that can be exploited by political interests, that will allow us to start building a system of artificial but effective warnings, warnings which will parallel the instincts of animals who flee before the hurricane, pile up a larger store of nuts before a severe winter, or of caterpillars who respond to impending climatic changes by growing thicker coats.

Human beings need protection as much as other animals, but they have no built-in ways of protecting themselves against the nexus between their great artificial human settlements, their worldwide system of intercommunication, and the enormous size of their interventions. As they harness atomic fission and invent electronic controls, miniaturization, computers, satellites, etc., which have woven the previously dispersed and unconnected populations of the planet into a single, interconnected, mutually dependent and totally-at-hazard single group, there is an urgent need for new manmade inventions by which the scientific knowledge which has made the new kind of society possible can interact with forms of political decisionmaking to make that single planetary community less vulnerable and more viable.

At the center of this problem lie the relationships between scientists, technologists, human scientists, and political decisionmakers. Inevitably, different political interests will seize upon disagreements among scientists to buttress their own interests and to discredit scientific advice. Scientists themselves may value making a fine point against a rival more than the possible consequences of the intra-scientific battle; or be extremely cautious so as to protect their reputations—among scientists—which is a modern equivalent of fiddling while Rome burns or dancing on the eve of the Battle of Waterloo; or they may simply despair of ever connecting effectively the nature of science, with its built-in requirement for validation by other scientists, into the political decisionmaking bureaucracies of the world. Very often scientists are content to blame either the politicians or the technologists who "exploit" the findings of "pure science," rather than making an effort to warn against the dangers of such shortsightedness and immediate exploitation.

There is as yet hardly any "law of the air" to parallel the age-old agreements about war and peace, ownership and trespass on land, access to streams and shores, and the more recent conventions governing the high seas. Most of the legal efforts to prevent pollution of the lower atmosphere have been localized in cities, or limited to state or provincial legislative efforts. Attempts to protect the atmosphere as a whole, with the recognition of the inevitable and unpreventable interconnectedness of all the peoples and all life itself upon the same earth-encircling atmosphere, are in their infancy.

It was not, however, accidental that Sweden, itself not only highly industrialized but also exposed to pollution from the Soviet Union, the United Kingdom, and West Germany, should have taken the initiative in convening an international conference on the protection of the environment. (An account of this is in Part 5.) This example demonstrates that we need a sober, cautious, credible statement of the extent to which pollutants, originating in one place on the planet, may affect neighboring peoples or all the peoples of a given region, or possibly even affect the life-carrying capacity of the total environment.

Only by making clear how physically interdependent are the people of all nations can we relate measures taken by one nation to measures taken by another in a way that will draw on the necessary capacities for sacrifice, dedication, and farsightedness of which human beings—as a group—have proved capable. We have witnessed their willingness to die if necessary and to suffer every sort of hardship to protect their children's future in their home countries. It is therefore the statement of major possibilities of danger which may overtake humankind—or all life on the planet—within the life span of those who are already born on which it is important to concentrate attention. The old impulse to deny danger (past or future), to continue to build our villages "on the side of the volcano," persists in the attitudes of the peoples of the world and in their leaders who seek support, compliance, and reelection. If irresponsible scientific controversies provide encouragement for these impulses, there is little hope of providing the future protection that is needed.

Another human impulse which we must attempt to overcome is the willingness to ignore danger when unidentified populations are at risk. There are a multitude of examples of this, such as our calm acceptance of tens of thousands of anonymous auto deaths each year contrasted with our unwillingness to contemplate the loss of a single astronaut. The question is how far we can continue calmly to accept the present degradation of parts of our environment and the resultant suffering and loss of human life, wherever it may be. There are, however, signs of a new awareness of populations in danger and an unwillingness to accept threats to the health of workers in chemical plants, for example, or to abandon people in hurricane-ravaged communities or starving nomads of the Sahel. Our ethic and our sense of responsibility are steadily widening from a concern with our own children and those of our neighbors to the children of other nations and to the peoples of the planet, but we have a long way to go before we can demonstrate that we are willing to act vigorously when nameless throngs are endangered.

It is important to realize that there is a human tendency to exaggerate a point in the heat of a debate and to overstate one's case in order to win an argument. Thus, those who react against prophets of doom, believing that there is not adequate scientific basis for their melancholy prophecies, tend to become in turn prophets of paradisiacal impossibilities, guaranteed utopias of technological bliss, or benign interventions on behalf of mankind that are none the less irrational just because they are couched as "rational." They express a kind of faith in the built-in human instinct for survival, or a faith in some magical technological panacea.

What we need to invent—as responsible scientists—are ways in which farsightedness can become a habit of the citizenry of the diverse peoples of this planet. This, of course, poses a set of technical problems for social scientists, but they are helpless without a highly articulate and responsible expression of position on the part of natural scientists. Only if natural scientists can develop ways of making their statements on the present state of danger credible to each other can we hope to make them credible (and understandable) to social scientists, politicians, and the citizenry.

This Conference is one step in that direction. As a human scientist concerned with institution-building for a newly interdependent planetary community, I have asked a group of atmospheric specialists to meet here to consider how the very real threats to humankind and life on this planet can be stated with credibility and persuasiveness before the present society of nations begins to enact laws of the air, or plan for "international environmental impact statements," or develop nationalistic barriers against intrusion on their "air space," or declare themselves independent of the consequences which their activities—with possible regional or global impact—may have on members of other nation states.

We have the experience of thousands of years of "international law" and its predecessors dealing with the land and several hundred years of the Law of the Sea behind us as precedents. However, we now have a totally new situation. Instead of dealing with a limited domain—for example, land on which two men cannot build their houses simultaneously or plant their crops, or areas of the seas in which fishing fleets compete—we now confront a domain that must either be shared and responsibly protected by all people or all people will suffer. The atmosphere is, in a sense, the ultimate international commons. To the extent that planetary hazards can be spelled out, we also have an unprecedented opportunity to provide a basis for wider responsibility than the world has ever known, evolved from a chaotic history that witnessed empires that crumbled and aggregations of mutually incompatible ethnic groups that united only to split apart again.

Those members of the Conference who are concerned with manmade mechanisms such as international law and city planning, or the development of new instruments such as environmental impact statements, civil aviation agreements, Earth Watch (ERTS), World Weather Watch, meteorological satellites, food banks, space exploration, and so forth, will be primarily asking questions, familiar as they are with the kind of arguments which are at present being pressed by the many sides of every controversy involving the impact of new technologies. The specialists in the atmospheric sciences and the technologies dealing with the impact of mankind's activities on the global atmosphere, on the air over cities, on the waters of the deep seas, and on the upper atmosphere will be seeking agreement on estimates of effects, couched in probabilistic terms, which can be presented so that internal scientific controversies cannot be used to blur the need for action—action either toward more adventurous interventions such as with earth satellites, or more watchfulness such as with ERTS, or more caution in the use of manmade chemical compounds or the harnessing of nuclear power.

We hope to arrive at a concise and persuasive statement directed specifically to problems of the atmosphere, anticipating and providing for scientific cooperation in the decisionmaking of the immediate future but having an effect on the far future.

New York, 1975 MARGARET MEAD

Part 1

Summary and Recommendations

Mankind has clearly demonstrated the ability to modify significantly the atmosphere of the earth. The following are examples of atmospheric changes that have either been measured directly or inferred indirectly:

- It has been observed that the carbon dioxide content of the entire atmosphere has been increased (and is still increasing) by burning fuels.
- It has been inferred that the oxides of nitrogen (NO_x) content of the stratosphere has been increased by high-flying aircraft, probably by extensive use of nitrogen fertilizers, and perhaps by certain air pollution abatement techniques that result in the production of nitrous oxide (N_2O).
- The chlorofluorocarbon content of the lower atmosphere has shown an observable increase due to the extensive worldwide use of these compounds as spray propellants and refrigerants.
- It has been inferred that the chlorine and hydrochloric acid (HCl) content of the stratosphere has been increased by the dissociation products of chlorofluorocarbons that diffuse upward.

Virtually all nations are contributing to these changes and all will be affected in some measure by their effects. Despite uncertainties inherent in estimating future consequences of manmade changes and difficulties in distinguishing the current changes from natural fluctuations, it is certain that indefinite growth in the level of human interference with natural processes will result in a number of clearly discernible effects. For example, increasing the content of carbon dioxide and chlorofluorocarbons in the atmosphere can enhance the "greenhouse effect," since both gases absorb infrared radiation. Model calculations suggest that the corresponding warming of average surface temperature from increases of both of these trace gases should be about 1°C by the year 2000; and furthermore the warming at high latitudes is expected to be many times larger. These estimates of warming are based on models that admittedly do not properly include all the important interactions in the climate system, and so could be in error by a factor of 2 or more; but, if they are indeed representative, the climatic warming that can be expected to occur in the next few decades from these global contaminants will be larger than any of the natural climatic fluctuations observed during the past 1,000 years or more. Whether this change would be generally beneficial or detrimental to society remains to be determined—the likelihood is that some regions of the earth will benefit while others will be hurt. Our current models of the climate system, models that include the atmosphere, oceans, land, and ice or snow masses, are not sufficiently complete to predict in detail the climate of a "warmer earth"—for example, the regional changes in precipitation and lengths of growing seasons.

Furthermore, future increases of oxides of nitrogen (NO_x) and of chlorine compounds in the stratosphere are expected to result in a decrease of stratospheric ozone (O_3), since in the presence of solar ultraviolet radiation both of these trace gases react with ozone in a catalytic cycle that destroys a certain fraction of the ozone, particularly at altitudes above 25 or 30 km. A decrease in the amount of ozone in the stratosphere permits more solar ultraviolet erythemal radiation (ultraviolet radiation capable of damaging essential molecules of living cells, notably DNA) to reach the surface, with consequences to living things that are still poorly determined but are certainly cause for concern. The expected increases of these trace gases in the stratosphere and the consequent reduction of ozone will occur gradually in the next decade and, if no remedial action were taken, probably would be noticeable by the year 2000 (estimates range from 5 percent to over 15 percent reduction of total ozone).

Model calculations indicate that a reduction of stratospheric ozone will result in a significant cooling of the upper stratosphere, but a relatively much smaller change of the mean temperature of the troposphere. The full climatic implications of an ozone decrease have not yet been spelled out.

A global change that, so far as we can tell, has produced no significant environmental effect is the buildup of the radioactive gases krypton-85 (^{85}Kr) and tritium (^{3}H) from nuclear power stations and fuel-processing plants in various parts of the world. Krypton-85 is a noble gas with a half-life of about 10 years, and there is no atmospheric removal process for it (other than its own radioactive decay). Tritium, which behaves like hydrogen chemically and has a slightly longer half-life, nearly always is incorporated in water vapor or liquid water. Its sink is the ocean, where it eventually is so diluted that its presence is of little consequence. Tritium can, however, be detrimental if it builds up locally in rivers or estuaries.

It has been suggested that, if nuclear power production continues to expand worldwide and no attempts are made to restrict the release of krypton-85 into the atmosphere, in about 50 years it will build up to the point where it may increase the conductivity of the troposphere by some 15 percent due to its ionizing radiation. A change in atmospheric conductivity could, it has been claimed, influence a variety of processes that probably depend on the natural electric field of the earth, such as the coalescence of droplets in clouds and the generation of lightning. Thus, the buildup of krypton-85 could influence the amount of precipitation, but this requires further investigation.

We have not dealt in any depth with the atmospheric changes that would be caused by further extensive testing of nuclear devices in the atmosphere, nor with the much larger effects of a full-scale nuclear weapons exchange. The former would add to the already existing background of radioactive debris in the stratosphere and on the surface of the earth and presumably would influence temporarily the ozone layer by creating some additional oxides of nitrogen. (Previous nuclear testing may have had a small effect, but this is not well established because it was superimposed on other natural fluctuations of total ozone.) An exchange of nuclear weapons between the major powers, on the other hand, would have so many disastrous consequences that the effects on the atmospheric environment would be *relatively* insignificant. It would be but one of many forms of self-inflicted harm to the survivors of the "conflict."

So far we have spoken of *global* effects of human activities, but changes of the atmospheric environment on a *regional* scale (an area which may nevertheless include several countries) due to the addition of manmade particles, heat, and chemical pollutants have already reached serious proportions in the industrialized parts of the world, and these effects know no political boundaries. A case in point is the observed increase in Scandinavia of "acid rain" containing significant amounts of sulfate resulting from the burning of sulfur-bearing coal and oil in other parts of Europe. Another example is the demonstrated change in rainfall downwind from large cities such as St. Louis, Chicago, and Paris.

A regional effect that has attracted wide attention in North Africa and elsewhere is the observed influence of forest clearing and of bad agricultural and grazing practices in marginal areas; and there is a theoretical conclusion that desert created or enlarged in this way will tend to remain a desert. It is very likely that the cumulative (and probably nonlinear) effects of these and other modifications of the character of the land already have played significant roles in regional climate change in many parts of the world.

As the extent of regional changes grows, the effect will be felt globally. Thus, the addition of heat from mankind's ever-increasing energy generation and release, and possibly the growing burden of airborne particles (aerosols) from industrialized areas and farmlands where slash-and-burn practices are followed, probably eventually will have an effect on the heat balance of the entire climate system and contribute further to the warming effects already discussed. The time scale for mankind's release of heat to have a significant global effect is estimated to be 50 to 100 years, depending on the long-term growth rate of energy demand and production and its geographical distribution.

A special aspect of the heat released from energy generation arises in connection with the likelihood of large "power parks," where immense amounts of heat must be dissipated over a limited area. The possible environmental problems associated with such power parks are already being studied.

The list of manmade changes of the atmospheric environment would not be complete without noting that the atmosphere is a transporter of many substances that do not remain in the atmosphere. Clouds of particles from industrial regions (most of them sulfate particles) are carried downwind, but only remain suspended in the lower atmosphere for about 5 days, on the average. A similar process carries such toxic substances as DDT, PCB's, heavy metals (including radioactive ones), and smog products across borders or over the ocean, where they are then deposited, primarily by fallout, rainout, or washout.

For all these reasons, it is our conviction that the atmosphere is a global resource whose preservation is the responsibility of all countries. National interests in the atmosphere already are becoming increasingly indistinguishable from global interests.

Furthermore, natural (or manmade) fluctuations in the climate will continue to have an impact on food production and, especially in marginally productive areas, there inevitably will be periods of markedly lower crop yields. Such fluctuations, regardless of their cause, clearly must be taken into account in both short- and long-range planning for society, in the face of growing populations and finite resources. While we are not optimistic that much skill in forecasting climate fluctuations a season or more in advance will be acquired in the near future, especially for middle latitudes, we note that an effort to keep a close watch on the climate of all parts of the globe would give some advance warning of crop failures and abundances.

It is with these thoughts in mind that we turn to the question of possible solutions to the problems raised, recognizing that serious decisions are now being made on the proper courses of action to avoid the damage to society that could accrue from changes in the atmospheric environment in the decades ahead. (To ignore the possibility of such changes is, in effect, a *decision not to act.*) These difficult decisions of necessity must be made on the basis of incomplete knowledge, since in many cases the time for significant changes to occur appears to be shorter than that required to develop a satisfactory scientific understanding of the major factors involved. It is therefore essential that policy makers and scientists work together continuously to make best use of our ever-growing but still limited information, and that they learn to communicate across the gap that often lies between the realms of politics and science.

To provide a framework that will assist in this decisionmaking process, we propose and urge that international agreements and mechanisms concerning the atmosphere be established (or evolved from existing ones) in order to:

- Exchange information.
- Warn of impending climate-related disaster, such as regional crop failures, by (for example) a continuously updated forecast of crop yields and inventory of food reserves.
- Develop strategies to help governments cope with such disasters, and where possible mitigate their effects *before* they occur.
- Foster research on the interdisciplinary problems relating to the interface between society and the natural environment.
- Ensure the timely application of science and technology of the atmosphere for the betterment of all mankind, with careful attention to both direct and indirect effects.
- Encourage monitoring of manmade and naturally induced changes of the atmospheric environment.
- Provide guidelines controlling purposeful weather modification activities, peaceful or otherwise.
- Govern large-scale operations or experiments on the atmosphere-ocean-ice-land system by one country or group of countries that may cause regional or global changes, and make that country or those countries responsible for damage caused by these changes.

In implementing this proposed framework of international agreements and mechanisms, it must be emphasized again that for many years to come there will be a limit on the confidence with which decisions can be made, a limit imposed by our lack of ability to make adequate predictions of the effects of natural or manmade changes of the environment. This inadequacy probably will be felt most explicitly in connection with the last two items, dealing with the effects of weather or climate modification activities. Nevertheless, the decisions have to be made eventually, and we believe that the development of a better scientific understanding and prediction ability should proceed together with the development of the international framework in which they can be applied.

We have spoken here of truly international (or supranational) efforts since we are dealing with a *global* problem. Nevertheless, the experience of international movements and organizations that have had similar objectives suggests that in some instances *regional* mechanisms are more likely to succeed at the outset since groups of neighboring countries tend to have somewhat unified viewpoints and may be motivated to work toward a common objective. Western Europe, East Africa, North America, and South America are examples of regional groups of countries that have already established some mechanisms for working together, and conflicts arising from air pollution or weather modification across borders are settled appropriately on a regional basis. In the long run, however, the global nature of many of the problems must be recognized if any satisfactory solutions to the most serious ones are to be found.

In this report we have used as examples of the increasing influence of mankind on the atmospheric environment those effects that have already been recognized as significant. Some of these effects have been recognized only for a few years—one has come to light just in the past year—so we are painfully aware of the possibility that even before this report is published some new or hitherto overlooked issue will surface, one that should have deserved the attention of the Conference.

Due to the extraordinarily complex nature of the interactions within the climate system, and between the atmosphere and society, it is essential that investigators pursue all facets of the questions raised in this report. It is certain that scientific curiosity will uncover new surprises and shed more light on recognized problems, but the nature of such inquiries is not always compatible with the standard compartmentalized structure of universities and most other research organizations. Thus, imaginative support of interdisciplinary and sometimes unconventional research must be encouraged, since the subject demands it—and time is already becoming a critical factor.

Part 2

The Atmosphere and Its Climate

ANTHONY BRODERICK, *Rapporteur*

THE LOWER ATMOSPHERE

Introduction

The approach taken to the broad subject being addressed by the conference was to start with a discussion of current understanding of the atmosphere and the "climate system." It was felt that this would be necessary for such an interdisciplinary group, so that those who were not physical scientists would have some acquaintance with the physical concepts involved. Perhaps even more important would be a clarification of the extent of our knowledge of these concepts, since a sensible discourse can take place only when the limits of uncertainty are understood.

In the following sections we review the pertinent facts about the atmosphere as we understand them, attempt to define the main factors that determine the climate, and discuss some of the things that mankind can do to affect it. These ideas also are extended to their societal implications under Regional Climate Interdependence.

There are several appendixes, however, that will be of special interest to specialists and nonspecialists alike. We call attention in particular to the reviews of *The Carbon Cycle and the Paleo-Climatic Record,* by Wallace S. Broecker; and *The Interaction of the Atmosphere and the Biosphere,* by James E. Lovelock.

Factors Governing Climate

"Climate" is usually taken to mean the average value (over a specified interval greater than a few weeks) of weather events—temperature, precipitation, winds, and so forth—and the statistics that describe their variations. The primary factors governing the earth's climate include the amount of solar radiation received, the earth's rotational speed, the composition of the atmosphere, the earth-surface properties and those of our oceans, and interactions among all of these. About 30% of the incoming solar radiation is reflected directly back to space (i.e., the earth "albedo" is about 0.3) by clouds for the most part, but also to a lesser extent by atmospheric molecules, dust particles, water droplets, and the surface itself. The majority of incoming solar energy is absorbed by the earth's surface and lower atmosphere.

Under conditions of constant solar radiation, and averaging over periods longer than a year, the amount of energy absorbed by the ocean–atmosphere–land system must be in balance with the amount of thermal infrared radiation emitted to outer space, or else the system's temperature would not be stable. While this near balance is achieved on a long-term global average, such is definitely not the case locally in time or space. Solar heating and cooling are highly variable both horizontally and vertically throughout the system. Tropical regions receive more solar energy per unit area than the poles. Tropical air thus heated rises, moving poleward, while colder air moves equatorward and downward to take its place. Thus, we have defined a simple "direct-cell heat engine," driven by the north–south temperature difference.

The rotation of the earth causes the mid-latitude air which originated near the equator to travel faster than that from the polar regions. This "coriolis effect" accounts for the westerly winds at mid-latitudes in both hemispheres, as well as the easterly trade winds near the equator. The north–south temperature gradient determines the strength of these winds, and the stronger this gradient, the more heat and energy are transported poleward. Large-scale storm systems (transient eddies) represent a self-stabilizing ("negative feedback") mechanism in the atmosphere to reduce the temperature gradient arising from the imbalance of solar heating, since they serve to transport heat poleward (Thompson 1961). As is well known, these storm systems are more vigorous and persistent in winter, when the temperature gradient is stronger, than in summer.

Heat transport in the atmosphere takes place in two forms. "Sensible" heat transport involves, for example, direct motion of warm air into a cold region. Transport of "latent" heat involves water vapor which is evaporated at the earth's surface, after which it can be transported to a higher latitude. Then, in the presence of suitable condensation nuclei (particles), the water will condense into droplets. During condensation, the "latent" heat originally required to convert the water from liquid to vapor is released. The process of evaporation, atmospheric transport of vapor, condensation, precipitation, and reevaporation is called the "hydrologic cycle," and is responsible for 25 to 30 percent of the mid-latitude heat transport from equatorial to polar regions. Sensible heat transport accounts for a like amount, and the remaining 30 percent or so is transported poleward by ocean currents.

Substantial areas of the earth are covered by ice and snow—the "cryosphere." In contrast to the stabilizing negative feedback of the hydrologic cycle between temperature and solar radiation, "positive feedback" is suspected in the case of ice and snow. Since they are highly reflective, if the surface area of the cryosphere were reduced, one would expect temperatures to increase (because more solar energy would be absorbed by the uncovered land and oceans), which would in turn stimulate greater melting, resulting in still higher temperatures, and so on (so-called "ice–albedo–temperature feedback"). The cryosphere, however, represents only a small fraction of the earth's surface. As mentioned above, clouds are by far the dominant reflectors of solar energy (Schneider 1972).

Thus, the hydrologic cycle is a major factor in determining climate through its influence on clouds, snow, ice, soil moisture, and surface vegetation. These processes, in turn, directly link atmospheric motions and ocean currents to the solar-thermal radiation balance and indirectly link them through the dependence of thermal radiation amount on the temperature of the radiator. Schematically, these complex climate-determining linkages are shown in Fig. 1.

Regional Climatic Interdependence

In 1968 the center of Columbia, Maryland, was neither warmer nor cooler than its surroundings. By the end of 1974, on a calm, clear day, the early evening temperature there was over 4°C warmer than its surroundings. By the mid-to-late 1980s, the temperature increase from outside the city to its center is expected to be about 6°C (Landsberg 1974, 1975). Why? In 1968 Columbia was a small crossroads, with a population of about 200 people. By 1974 urban development had created a city with a population of 20,000, and it is expected to reach 100,000 within a decade or so. The temperature gradient is a manifestation of the urban "heat island," a measured, local climatic change.

Larger metropolitan complexes are known to result in more than merely local climatic effects. For example, a city is a source of condensation nuclei from mobile and stationary combustion sources. These small particles are transported upward by warm air and perhaps into passing clouds of water vapor, where they can initiate condensation. From measurements in St. Louis, it appears that presently some areas 10

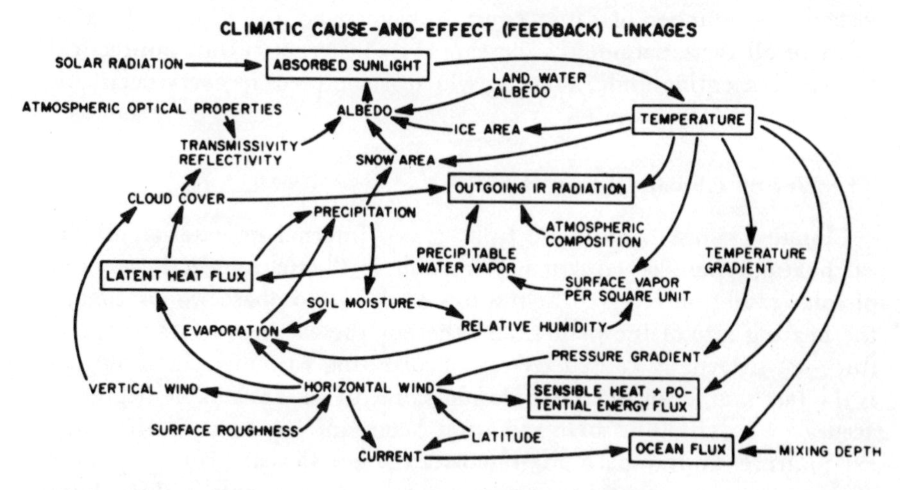

FIGURE 1. Climatic cause-and-effect (feedback) linkages. This diagram by Sellers shows the many cause-and-effect linkages that must be taken into account in a comprehensive climatic model (Kellogg and Schneider 1974). Reprinted with permission of *Science*.

to 15 miles downwind receive about 10 to 15 percent more rain than was normal a few decades ago (Landsberg 1975, Dettweiler and Changnon 1976). Since the city of St. Louis did not supply this added rainfall, who was "robbed" of their share? It is not yet clear, but someone may be the loser in this regional perturbation.

Industrial complexes on large scales are now evidently affecting considerably more distant areas, across the political boundaries of one or even several nations. Measurements of the concentrations of photochemical oxidant (ozone) and trichlorofluoromethane over the United Kingdom by Cox et al. (1975) have shown anomalously high concentrations of these pollutants. Back-tracing the air trajectories led them to conclude that this material originated in northwestern Europe, about 1,000 km distant, and that, on occasion, continental European emissions contribute significantly to photochemical smog pollution in the United Kingdom.

Acid rain is another environmental pollutant whose source may be outside the nation which feels its effects. The work of Brosset (1975a, b) shows clearly that acidification of land and lakes in the Scandinavian countries, as well as airborne ammonium sulfate particulate pollution, is caused by air from the European continent. These pollutants originate in both Eastern and Western European nations.

Clearly, then, we need not speculate: man *has* demonstrated the capability to modify the environment on an international, albeit regional, scale. As we examine the evidence, it can be seen that the scale of the effect increases as the size of the system under study is increased. But along with this observation goes the realization that the complexity of the system, the number of cause–effect linkages, and the degree of variability of all these parameters increase even faster. It is thus impractical to expect scientific understanding which is complete in every detail for global-scale climatic studies.

Theories of Global Climate Change

Climate change could arise from causes internal or external to the earth–atmosphere–ocean system. For example, fluctuations in the amount of solar radiation received at the top of the atmosphere would change the heating rate of the planet, and thereby the earth's surface temperature—an externally caused change. Frustrating to many climatologists is the fact that, even today, we do not know to any reasonable degree of accuracy the absolute solar radiation "constant," even with the large expenditures on the space program over the last 15 years. Not only is the absolute amount of total incoming radiation known only within about 1 percent but the spectral distribution of the solar radiation is even less well established. It is within the realm of reasonable possibility that all climate changes of the past could be explained by small, otherwise unnoticed fluctuations (for unknown reasons) in the sun's intensity of 1 percent or so.

Another possible external cause of climate change is fluctuation in the earth's orbital characteristics and the direction of its axis of rotation. Such variations, at intervals of 10,000 to 100,000 years, have been postulated to be the cause of past ice ages, and the climatic record tends to support this theory.

Changes in atmospheric dust and/or carbon dioxide content or changed land surface characteristics (e.g., desertification) are partially internal to the planetary system. Man's activities (or volcanoes) could effect such changes "externally," but variations in the planetary climatological state could have similar effects: winds (transporting dust), rainfall (affecting vegetation), temperature (affecting carbon dioxide's solubility in ocean water).

Of late, several internal causes of climate change have been postulated. Quasi-periodic or anomalous fluctuations in the ocean-surface temperature pattern could be one such cause (Namias 1972). Decreases in salinity of the North Atlantic or Arctic oceans, leading to increased sea ice formation, represent a second possible cause (SMIC 1971). Lorentz (1970) has proposed that in a system as complex as the earth–ocean–atmosphere one could have long-period self-fluctuations even with fixed external inputs—the so-called "almost-intransitivity" of the system.

Clearly, when climate does change (and we know from paleoclimatological and other records that it has), sorting out the proximate cause is a difficult task. More important to the purpose of this discussion, however, we must recognize that long-term "natural" climatic fluctuations must be separated from two other aspects of the problem: short-term fluctuations in weather (e.g., the "year without summer" of 1816) and man's activities. It is important to recognize that short-term weather anomalies are as much a property of the climatological state as the long-term average of the same variable. Attempts to understand these various factors, attribute cause and effect, and estimate the important characteristics of future climates require a quantitative description of the system behavior. While we presently do not have a completely satisfactory theory of climate change to use in this process, many of the important factors are reasonably well understood. The mathematical representations of our best current understanding take the form of computer-based climate "models" of varying degrees of sophistication.

Climate Modeling

It can be demonstrated (e.g., Leith 1971; Ramage 1976; Robinson 1975) that the complexity of the climate system is such that it is not possible to predict even with reasonable accuracy the value of a climatic variable (wind, rain, temperature, etc.) on an exact day more than, at most, 2 to 4 weeks in advance. Such detailed predictions require specific determination of future change in a fluid dynamic system (the atmosphere) which is initially specified on some microscopic scale of space and time. One could not specify the initial instantaneous position, velocity, and momentum of each and every molecule. This loss of detail in performing the initial averaging process results in an unspecified error. This initial error is not reduced as we project further into the future but, in effect, grows so that useful information cannot be obtained from the predictive equations at some time in the relatively near future. For large-scale eddies (high and low pressure systems), the time period over which predictions eventually may be useful is on the order of one week. For smaller scales such as a squall line, the time is reduced to only about an hour.

But our inability to perform such a detailed prediction does *not* necessarily mean that attempts to predict future climates on a more gross scale will be unsuccessful. It may be possible to develop a climate model useful for predicting, say, the current time rate of change of surface temperature (prediction of the first kind) or the change in long-term-average equilibrium surface temperature in future years (predictions of the second kind). Developing such a capability is one of the principal aims of current efforts in the field.

Climate models employ mathematical expressions of basic physical laws: conservation of mass, momentum, and energy. The complexity of the system feedback linkages (see Fig. 1) and the scale of the system under study, along with size and speed limitations of available computers, force the scientist to omit explicit treatment of many small-scale processes. For example, though cloudiness is of obvious importance as a climatic variable, models cannot be constructed which treat the development of individual clouds. In this and similar situations, what is done is to attempt the development of a physically meaningful ensemble average based on realistic assumptions about the statistical properties of the variable. Such a treatment is called a "parameterization." It is the validity of these prescriptions for parameterization which form the "canons of faith" upon which the art of climate modeling is presently based.

Since presently available climate models cannot include all of the sometimes competing variables which affect the climate system as a whole, it is necessary to select those variables thought to be of primary importance to a particular problem. One important consideration is the time scale under consideration. Except on regional scales, for example, land-sea interactions are thought to be of relatively minor importance for time scales of the order of a few years or more. The large heat capacity of the deep ocean, on the other hand, is believed to make such interactions with it important over long time periods (hundreds to thousands of years).

Inevitably, in a model of the climate system some feedback loops which may be of importance might be ignored. This may be deliberate, due to computational limitations, or inadvertent, due to ignorance of their existence. Under these conditions, the ultimate effect on the reliability of a model prediction is not clear. For example, consider the case of a doubling of atmospheric carbon dioxide concentration. Attempts to model the effects of such a change over the past decade or so have resulted in extreme predictions of an average global surface temperature warming of 0.7°C (Weare and Snell 1974) to 9.6°C (Möller 1963), with a generally agreed "reasonable" best estimate of about 2.5°C. (This matter will be discussed further under Inadvertent Climate Modification and Appendix I.) The reasons for the large spread in estimates have been analyzed and explained by Schneider (1975). What about some of the effects which need to be parameterized better? Such a warming would tend to melt snow and ice, reducing the surface area of the cryosphere which would lower the amount of reflected solar energy resulting in more heat absorption and even higher temperatures. (This feedback currently is included in the better climate models.) *But*, this warming also would tend to result in more water being evaporated, which could increase cloudiness and correspondingly increase the amount of reflected energy, thereby counteracting the warming trend. *But*, if the cloudiness increases, that would tend to increase trapped thermal infrared radiation, causing a warming trend. This recitation of links in the complete chain can go on for a long time. The point to be made is that the complexity of the system is such that one cannot a priori expect that these other factors will tend either to reduce *or* enhance the effect predicted by a simpler representation.

Approaches to Modeling and Validation

A climate model is defined by a particular choice of meridional and zonal resolution (the smallest "boxes" in the horizontal grid), vertical resolution, parameterization schemes, time resolution, and numerical method of approximation to the fundamental fluid dynamics equations. The simplest type of climate model is one which averages horizontally over the entire globe, but includes a variation with altitude of important (globally averaged) atmospheric properties. Such a one-dimensional model usually prescribes the relationship between temperature decrease with height (in the troposphere) and vertical convection and includes the variation in density of important atmospheric trace gases and particles with altitude. They are useful tools, and give some indication of the relative importance of concentration of individual atmospheric constituents to maintaining the balance between solar and infrared radiation. The work of Manabe and Wetherald (1967) with such a model is cited widely as yielding a reasonable estimate for the effects of increased atmospheric carbon dioxide content. Such estimates must be viewed with some skepticism, however, since, for example, the very fact that solar heating is not actually uniform over the surface of the earth is a primary reason for atmospheric behavior. The latitudinal variation of solar heating, of course, cannot be treated explicitly in such a one-dimensional model. This is not to say that such an estimate will be, ipso facto, substantially in error, but only to point out that any accuracy it may turn out to have is the result of a fortuitous choice of approach which is not entirely defensible from a physical viewpoint.

Zonally averaged models represent the second upward step in terms of model complexity. Here the energy balance equations are explicitly treated only in specified zonal bands (perhaps 5° or 10° wide), and horizontal transport is parameterized in one of several semi-empirical schemes that relate heat transport to mean latitudinal temperature gradient. Including the previously mentioned ice–albedo–temperature feedback relationship to surface temperature, Budyko (1972, 1974) and Sellers (1969, 1973) developed such a model, which has a remarkable sensitivity to changes in the solar energy input. A decrease of only about 1 percent in the solar radiation is predicted by these models to lead to the kind of ice age that existed 20,000 years ago. An increase of similar magnitude is predicted by them to melt the polar icecaps. Schneider and Gal-Chen (1973) and others have given additional information on the utility and behavior of such models.

It is of interest to note how different modeling approaches may result in differently perceived climatic effects from the same type of system perturbation. Again, we can draw the conclusion that inclusion of additional feedback mechanisms does not necessarily improve the validity of a model prediction, and could in principle even reverse the sign of a predicted climate change (though this is unlikely).

The most complex atmospheric models in use today and those which provide our best hope for developing physically defensible prediction methodologies are the general circulation models of the atmosphere (GCM's). These constructs represent an attempt to minimize the use of parameterization in order to base the model, as much as possible, on fundamental physical relationships. Such a model explicitly treats atmospheric motions in all three dimensions, with a present resolution of perhaps 250 km in the horizontal at perhaps 25 atmospheric "levels" spaced between the surface and the model "top" (usually in the stratosphere or lower mesosphere). To be ultimately successful, the ocean-land system interactions also must be treated explicitly in such a model, but they have not been included properly in any of the GCM's to date. Perhaps the most elaborate modeling effort to date, a GCM with an interacted but noncirculating ocean, is that of Manabe and Wetherald (1975) which has been applied with good results to the CO_2 doubling problem.

As the amount of information being modeled increases, severe strain is imposed on computation requirements. This has led some to suggest that the GCM's probably will not prove to be practical tools for long-term climate forecasting for many years to come (Kellogg and Schneider 1974). It is suggested that perhaps some type of "compromise" will prove desirable, such as using a very coarse three-dimensional grid with extensive parameterization of subgrid-scale processes (a statistical-dynamical model), in order to cut down on computational requirements.

An important consideration here, however, is the inevitable requirement for model validation. One might suggest "validating" the statistical-dynamical approach by comparison with a high resolution GCM, but this does not obviate the requirement to compare one (or the other) to reality. It is well known that we do not have a detailed climatic record going back more than a few decades. Even this record is abysmal for its lack of information about the climatic conditions almost anywhere outside the developed countries, particularly over that 75 percent of the earth's surface that is ocean. Oceanic datum itself is almost nonexistent even today (e.g., sea-surface temperatures, *in-situ* temperatures below the surface, upwelling characteristics, etc.). Shannon's sampling theorem places constraints on our ability to validate any model: if we want to describe accurately a time series, it must be observed at a minimum frequency of twice its bandwidth. Thus, the effects of the 11-year solar cycle will take at least 22 years to define. But Dansgaard et al. (1971) have documented fluctuations in air temperature, as reflected in Greenland ice cores from the Camp Century site, of $\pm 1°C$ with significant energy at frequencies of 80 and 180 years (using power-spectral analysis). If these are indicative of global temperature fluctuations (and it is not clear why they should necessarily be so), any model attempting to simulate these phenomena will wait a long time for validation.

Only a few institutions around the world have made a serious commitment to the development of climate models useful for predicting natural climate changes. Their job is a difficult one which will take many years of hard effort before coming to fruition—and some are pessimistic about the outcome.

Inadvertent Climate Modification

In general, it now is accepted by the scientific community that man has the ability to act in such a way, purposefully or inadvertently, to modify climate on a global scale. He can do so by directly or indirectly modifying the heat balance of the planet. Earlier, the local climatic effects of the urban heat island were mentioned. To the list of potential local-climate modifiers we can add large-scale desert irrigation, defores-

tation, and desertification from overgrazing (all of which alter the surface reflectivity); damming of rivers to create large artificial lakes, damming the Bering Strait or eliminating the Arctic Sea ice pack (which modify the hydrologic cycle and, in some cases, surface reflectivity); and slash-and-burn clearing or otherwise adding large quantities of dust and smoke to the atmosphere (which would alter its albedo characteristics) (SMIC 1971). Some of these are illustrated schematically in Fig. 2.

One of the most well-studied hypotheses of human intervention in the climate system involves alteration of the amount of various optically important trace constituents of the atmosphere. These are the gases which absorb some of the thermal infrared radiation emitted by the ground, thereby acting as a thermal regulator of sorts by controlling the amount of heat radiated to outer space and maintaining the surface at an attractive temperature for life as we know it (Schneider and Kellogg 1973.) By changing the amount of naturally present, optically important gases such as carbon dioxide (CO_2), water vapor (H_2O), and ozone (O_3), we would seem to be able to alter atmospheric heat balance and, thus, the surface temperature of the earth (see Fig. 3). Similarly, by adding significant amounts of certain infrared-absorbing gases which are not naturally present in substantial quantity, we also run the risk of modifying the surface temperature. Ramanathan (1975) recently has pointed out that such might be the case with chlorofluorocarbons.

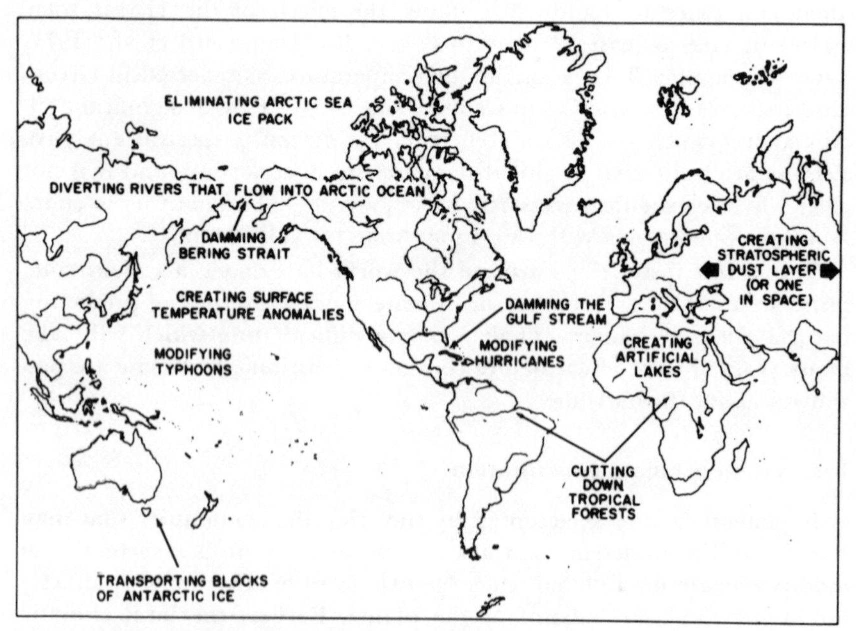

FIGURE 2. Some suggestions for inadvertent climate modification (Kellogg and Schneider 1974). Reprinted with permission of *Science*.

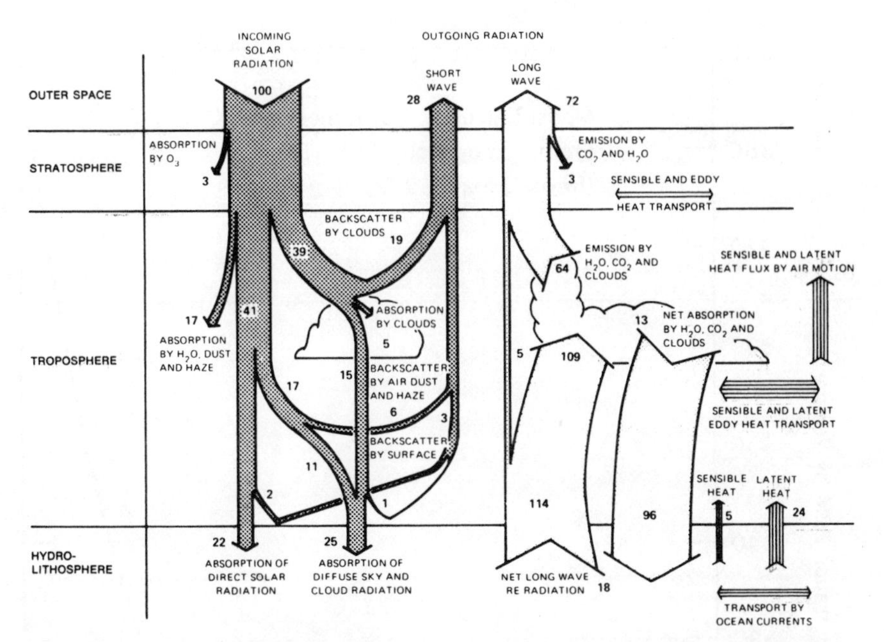

FIGURE 3. This figure, from Rotty and Mitchell (1974), shows the balance between incoming (solar) radiant energy and outgoing (terrestrial infrared) radiant energy, and the distribution of energy in the global system. Any disruption to these flow rates by human activities (e.g., energy production) has the potential to affect the earth's climate. (Also reprinted in Schneider 1976.) Reprinted with permission of Oak Ridge Associated Universities.

Since the industrial revolution, man has burned fossil fuels at an increasingly rapid rate due to the combination of population growth and increased per capita energy consumption. Figure 4 shows one tangible result: a buildup of atmospheric carbon dioxide concentration. Apparently, a little less than half of this gas has been absorbed into the oceans and forests, while the rest remains in the atmosphere. Recent estimates (e.g., SMIC 1971; Machta and Telegadas 1974) place the atmospheric CO_2 concentration of the early 19th century just above 290 parts per million (ppm) by volume. Since then, we have seen an increase of about 10 percent in roughly 110 years. But Fig. 4 indicates a projected time of about 18 years for the second 10 percent increase, and only 7 years for the third. Thus, while natural climate-stabilizing mechanisms have apparently dealt with a change of 1 percent every 10 years, it is reasonable to ask if the climate system will respond in the same way to a CO_2 increase of 1 percent every 2 years.

In parallel with adding carbon dioxide, one would expect addition of aerosol particles from industrialization, increased agricultural burning, and other sources. However, documentation of such a global increase has not been done. What appears to be the case (Kellogg and

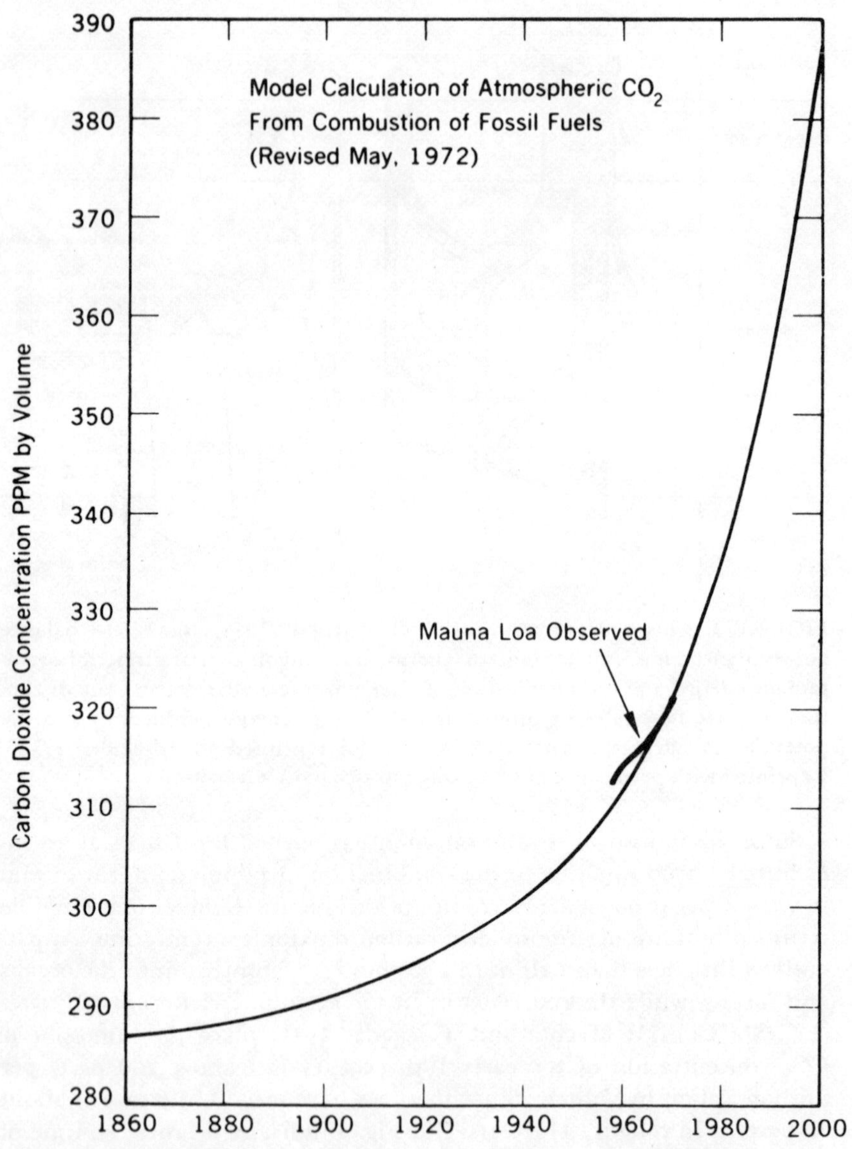

FIGURE 4. A calculation of the growth of atmospheric carbon dioxide concentration, in parts per million by volume, from the Industrial Revolution to the end of the 20th century. In this model it is assumed that the current 4 percent per year growth rate of fossil fuel combustion would be followed by a 3.5 percent growth rate after 1979. Takeup by the world's oceans and the biosphere are taken into account. The short segment from 1958 to 1971 represents the observed concentration at the Mauna Loa Observatory. (Taken from Machta, U.S. Atomic Energy Commission Report, 1973).

Schneider 1974; Kellogg et al. 1975) is that increases in local aerosol concentrations, downwind of the source, are in evidence. Over the oceans, such an increase would be expected to have a cooling effect since the reflectivity of the ocean-atmosphere system would be enhanced by atmospheric particles. Over land or snow surfaces, however, quite the opposite is expected, since the reflectivity of most particles is probably lower than that of the surface, and a net warming might result. Bryson (1974), Mitchell (1971), and Kellogg et al. (1975) have analyzed this particular perturbation from three points of view. On balance, the question of whether warming or cooling might be expected from increased anthropogenic aerosol concentrations has not yet been answered definitively, but it is most likely to cause a warming.

When reviewing possible effects on the heat balance of the planet, one also might ask if the direct heating effect of increased energy use is of significance. At present, mankind's total energy consumption and heat release represents only about 0.01 percent of the average solar power (155 watts per square meter) absorbed at the surface. Man presently represents, therefore, only a small part of the global energy budget, as schematically illustrated in Fig. 3.

Projections into the next century or two, however, show that man's influence conceivably could become significant on a global scale (Schneider and Dennett 1975; Kellogg 1974, 1975; Broecker 1975). Regional nuclear power "parks," with densities of 10 to 40 thousands watts per square meter, could produce severe local climatic disruptions. (Reconnaissance photographs taken downwind of active volcanic sites in Iceland have shown tornadoes which were apparently induced by volcanic eruptions.)

How Confident Are We in Our Models?

So far, we have produced a long list of human actions that possibly could modify the global climate. We also have reviewed briefly the status of climate models which must be employed to analyze the potential implications of these actions. In the course of the discussion, it was emphasized that man already has demonstrated the ability to modify climate on a regional scale, with potential effects which recognize no political boundaries. It appears that human activity has reached a scale and rate of growth which makes it possible to act inadvertently or purposely to modify climate on a global scale.

At the same time, it is clear that the system of feedback linkages between cause and effect which characterizes the climate system is exceedingly complex. Our understanding of this system is relatively primitive, and our ability to model its behavior is quite limited. What, then, can the scientific community say which will be useful for those in a position to make important decisions—by their very nature, political—in na-

tional governments and international forums? There are, at present, two schools of thought on this crucial issue.

As a simple example, look at the issue of increased carbon dioxide concentration resulting from a continued increased use of fossil fuels. Simplified climatic "models" lead one to expect a surface temperature warming from such activity. But these models do not treat properly all of the known compensating effects which could reduce such a warming, let alone those which have not yet been identified. On the other hand, a careful study of the complex feedbacks which characterize the climate system may uncover similarly unaccounted-for effects which would tend to enhance the warming. Briefly stated, the system cannot be completely characterized, and we cannot *categorically* state that the CO_2 effect predicted using our best models is going to be greater or less than the change which will actually occur.

One school of thought insists that there is technical agreement that increased CO_2 concentration by itself, *all other things being equal,* will lead to global warming. Admitting that there are myriad other related effects, the argument is presented that they are probably relatively small, and that in any case there is an equal likelihood that these unaccounted-for effects would either enhance or reduce the calculated effect. It is concluded that, in cases where the societal risk is great, one should therefore act as if the unaccounted-for effects had been included, since we have no way of dismissing the very real possibility that the calculated effect will prevail.

The other school begins with the caution that there has never been a global climatic change whose explanation has been accepted by the scientific community as a whole. Recognizing the necessity for using climate models and simulation techniques, this group may be characterized by the following, highly conservative statement: "If current physically comprehensive models are inadequate to answer some of our questions, then certainly we should be wary of basing broad national or international decisions on hand-waving arguments or back-of-the-envelope calculations." (Smagorinsky 1974).

While the opinions expressed above may appear to provide two distinctly different recommended courses of action (the second being *no action* at all for the time being), adherents to either philosophy are in general agreement on the need for attention to these matters. It is essential that serious consideration be given these issues now, and that scientists and policymakers develop a close working relationship to optimize the use of our rapidly growing understanding of the subject of mankind's impact on climate.

MAN'S INFLUENCE ON STRATOSPHERIC OZONE

Atmospheric Structure

The atmosphere is divided conveniently into concentric shells, considered more or less distinct because of the different change of temperature with increasing altitude in each. Starting at ground level, temperature decreases with height to a first minimum at the "tropopause." The region between the surface and the tropopause is called the "troposphere"; it extends to a height of about 8 km in polar latitudes, reaching to about 16 km near the equator, and contains roughly 85 percent of the total atmospheric mass. The troposphere is well mixed, due to turbulent winds and large-scale convective motion, and the air is cleansed frequently of aerosol particles by the scavenging action of precipitation.

At altitudes just above the tropopause, temperature is roughly constant or slowly increasing with increasing altitude, then it increases quite rapidly to a secondary maximum at the "stratopause" at about 50 km. (At the stratopause the average temperature is roughly comparable to that at the surface.) This region of permanent thermal inversion is called the "stratosphere," which contains most of the remaining 15 percent of the atmospheric mass. The stratosphere is characterized by extreme vertical stability. The winds there are predominantly east-west, or zonal in direction, but there are slower meridional circulations. Thus, it is conceivable that a "puff" of pollutant injected into the stratosphere could be spread, after a few weeks, in a narrow zone roughly along latitude lines. Then this band would slowly expand in a north–south direction throughout the hemisphere of injection, and, after several months or more, the slow vertical exchange of the region would bring part of the pollutant down to the troposphere, where it would be rapidly removed by rainfall or turbulent mixing.

The stratosphere is thus a relatively "stagnant," stable atmospheric region which is susceptible to a buildup of pollution. Typically, "residence times" in the middle stratosphere at about 20 km (i.e., the time after injection of an inert tracer when half to two-thirds of the material would be removed to the troposphere) are considered to be on the order of 2 years. Because of such a long residence time pollution of the stratosphere can result in a truly global problem. Stratospheric pollution over the United States cannot be kept from affecting the rest of the Northern Hemisphere and, if severe enough, the Southern Hemisphere.

The Ozone Region of the Stratosphere

The warmth of the upper stratosphere is caused by the presence of ozone, which reaches a peak concentration near 25 km (but a peak mixing ratio—parts per million by volume—near 30 km). Ozone is a very minor constituent of the atmosphere (less than about 0.0001 percent), but is an extremely effective absorber of biologically active solar ultraviolet radiation (UV-β). By absorbing this radiation, the ozone heats the upper stratosphere, creating the thermal inversion characteristic of the region.

Ozone is present in the stratosphere in what might be called a dynamic equilibrium, the result of a complex array of competing chemical and transport actions. At high altitudes, in the principal ozone "source" region near 30–35 km, the ozone is formed by the dissociation of molecular oxygen and subsequent recombination of the atomic oxygen thus produced with molecular oxygen (in the presence of a third body):

$$O_2 + h_2\nu \rightarrow O + O$$
$$O + O_{+}M \rightarrow O_3 + M$$

Ozone is also destroyed by dissociation:

$$O_3 + h_2\nu \rightarrow O_2 + O$$
$$O_3 + O \rightarrow 2O_2$$

The above scheme is the classic Chapman (1930) set of reactions, which formed the basis of our understanding of ozone formation and destruction for several decades, until the mid-1960s.

To understand the behavior of ozone requires a detailed understanding of the balance among all the chemical reactions and dynamic motions which produce and transport ozone (its "sources") and the mechanisms of its destruction (the "sinks"). Only then can one derive the steady-state average (equilibrium) concentration of stratospheric ozone.

In the early 1960s several attempts to balance these reactions were deemed "successful" (e.g., Prabhakara 1963), but by the late 1960s more accurate analyses of the chemical reactions of importance showed atmospheric scientists that there were large gaps in their understanding of this matter. At that time attempts to balance the source and sink terms of the chemical–dynamic relationship were unsuccessful, since they indicated that nearly twice as much ozone was being produced per unit of time as was being destroyed. It was then that Crutzen (1970, 1974b) pointed out that naturally occurring oxides of nitrogen (NO_x) could be assigned the role of a major sink of ozone in the stratosphere, and that reasonable estimates of their concentration would result in a fairly satisfactory balance of the sources and sinks of ozone. Working independently, Johnston (1971) called attention to the potential danger of SST-engine oxides-of-nitrogen emissions which might upset the natural balance, and lead to a decrease in atmospheric ozone concentration.

Ozone Destruction by Nitric Oxide

Nitric oxide destroys ozone in a catalytic manner. Though the complete chemical reaction scheme is very complex, the most important reaction is as follows: a nitric oxide molecule first reacts with an ozone molecule, resulting in the production of one molecule of oxygen and one of nitrogen dioxide. The nitrogen dioxide molecule, in turn, combines with an oxygen atom (which could otherwise have formed an ozone molecule) to produce an oxygen molecule and a nitric oxide molecule. The cycle has then gone full circle—having destroyed two ozone molecules, the nitric oxide molecule remains to begin another ozone destruction cycle.

$$NO + O_3 \rightarrow O_2 + NO_2$$
$$NO_2 + O \rightarrow O_2 + NO$$

Thus, because of the "catalytic" nature of this chemical-reaction chain, and others of somewhat lesser importance, very small amounts of nitric oxide can destroy surprisingly large amounts of ozone, and will continue to do so until the nitric oxide or its derivative (nitric acid) is physically removed from the stratosphere by the natural process of slow, downward vertical transport to the troposphere, where it is scavenged by precipitation processes.

This is the essential understanding which was developed in the period 1971–75, spurred by the U.S. Department of Transportation's Climatic Impact Assessment Program (Grobecker et al. 1974) **and parallel efforts in other countries throughout the world. Stated simply, we now understand that the ozone concentration in the stratosphere is the result of a dynamic balance, involving a host of competing chemical and transport mechanisms of ozone production and loss. Certain pollutants (such as the oxides of nitrogen) can interact with ozone in a catalytic, self-perpetuating cycle of ozone destruction. Once injected into the stratosphere, any gaseous pollutant will remain there for a long time, due to the lack of rapid vertical movement in that region of the atmosphere, while it is spread horizontally over the hemisphere of injection. Therefore, if the stratosphere is subjected to pollution by materials which can catalytically destroy ozone, a surprisingly large destruction of ozone may occur on a global scale. For example, according to the figures of** Grobecker et al. (1974), **if the U.S. SST had been built in a quantity of 500 aircraft, and if these had cruised 7 or 8 hours per day in the stratosphere and emitted 18 grams of nitric oxide per kilogram of fuel burned, the resulting average ozone reduction is estimated at about 15 percent** (as a **Northern Hemisphere average**).

Other "Threats" to Ozone

Unfortunately, nitric oxide from high-flying aircraft is not the only pollutant which might reduce stratospheric ozone levels (Broderick 1976). **Other pollutants, notably hydrochloric acid (HCl), a dissociated product of certain chlorocarbons that diffuse upward from the tropo-**

sphere, are capable of destroying ozone in the same kind of catalytic process that we described for nitric oxide. The list is growing rapidly, and now includes the following items:

- Nitric oxide, from nuclear weapons tests or a nuclear war (National Academy of Sciences 1975).
- Nitric oxide, from the nitrous oxide thought to be added to the atmosphere by agricultural denitrification of nitrogen compounds in fertilizers (Crutzen 1974b; McElroy 1975).
- Chlorine, from the chlorofluoromethanes used as an aerosol spray propellent and as a refrigerant (Crutzen 1974a; Molina and Rowland 1974).
- Chlorine from the main booster engine of the NASA space shuttle (National Academy of Sciences 1975).
- Bromine, from methyl bromide used as an agricultural fumigant (McElroy 1975).
- Bromine, from brominated chlorocarbons used as a firefighting agent (Wofsy et al. 1975).

In addition, research is leading to discovery of natural causes of ozone destruction. For example, Ruderman and Chamberlain (1975) calculated that the 11-year cyclic variation in ozone concentration, which is as much as 5 percent near the poles may be the result of solar-induced modulation of galactic cosmic rays (which produce nitric oxide in the stratosphere). Crutzen et al. (1975) have shown, however, that the cosmic ray effect probably is not as great as that of solar flares which should be considered as potentially important sources of stratospheric nitric oxide because of the ionizing and dissociative action of fast secondary electrons which result from polar cap absorption of solar protons. Finally, Ruderman (1974) showed that nearby (within perhaps 50 light-years) supernova explosions could be capable of removing most of the world's ozone "cover" by action of their expected flux of ionizing radiation, and that such an event may be expected to occur every few hundred million years.

Obviously, there is little to be done to avoid purely natural occurrences of ozone reduction—certainly those which are completely external to our planetary system. As regards other causes, i.e., those which are essentially anthropogenic, we do have some control. But we must recognize that we probably do not yet understand fully all of the factors which affect ozone. The possibility does exist, therefore, that our present understanding, briefly outlined in the above paragraphs, is inaccurate, and that we are *not* on the brink of a serious global pollution incident (Lovelock 1974). Nitric oxide, chlorine, and bromine are all present in small amounts in the "natural," unperturbed atmosphere. The sources of these materials, and their sinks, are not yet understood fully (e.g., Johnston 1975; Lovelock 1974; McElroy et al. 1976).

Consequently, their natural flux into the stratosphere is not yet established with good accuracy. Clearly, if we have not established with accuracy the natural source strength of these pollutants, any calculation of the effects of anthropogenically increasing the strength of this stratospheric source will be at least of similar inaccuracy.

Effects of Reduced Ozone

As mentioned earlier, ozone is a powerful absorber of biologically active ultraviolet (UV-β) radiation. The ever-changing dynamic equilibrium of ozone concentration results in a variability in space and time which is quite large. Typically, the average annual ozone "total column" concentration (integrated from the ground up through the stratosphere) varies by as much as a factor of 2 from equator (low) to pole (highest). At any one place the day-to-day variability might be on the order of 25 percent, with similar cyclic seasonal-average variation (low in fall, high in spring). Thus, the amount of biologically active UV-β radiation reaching the earth's surface is variable, in direct (though nonlinear) relation to the variability of ozone, all other things (clouds, particle distribution, etc.) being equal. Indeed, using a "Robertson meter" (Robertson 1972) to measure it, the variation in the UV-β radiation over, say, Philadelphia has been observed to vary by about a factor of 6 over the year, with wide (more than a factor of 2) day-to-day variability (Sundararaman et al. 1975).

But wide day-to-day, seasonal, or yearly variability in ozone does not necessarily enable one to dismiss smaller changes in the *long-term average* ozone amounts—such as the 10 percent or more that some have calculated might result from anthropogenic perturbations. Even though ozone has wide variability in space and time, its average value is relatively stable over periods longer than a year (McDonald 1971). Thus, a small perturbation in the average ozone concentration might not be detectable, but would, in fact, result in an increase in the total UV-β dose received at the surface over a long period of time.

The biological effects of such a change are far from well established. In the past few years, considerable attention has been focused on the projected increase in skin cancer which might result from increased average UV-β levels (e.g., Grobecker et al. 1974; National Academy of Sciences 1975). There has also been speculation that other adverse effects might accrue from increased UV-β dosages—effects on plants and animals and even decreased useful life of certain paints and materials due to increased weathering (IMOS 1975). However, the fact that most of the claimed adverse effects (even logically expected to occur) are associated plausibly with increases in UV-β does not ameliorate the lack of an accepted causal mechanism for such effects.

In all cases, it is clear that considerably more work needs to be com-

pleted before reasonably accurate relationships may be derived between increased UV-β and adverse biological effects. Most of these adverse effects are of sufficiently significant consequence to require that we move cautiously at present, under the assumption that some, if not all, will be demonstrated to be linked mechanistically to UV-β. The extent of the actions which should be taken now to avoid such potential risks is a difficult matter of public policy, which must be developed by a close cooperation of the scientific and political communities.

Part 3

Human Costs and Benefits of Environmental Change: Limits to Growth in a World of Uncertainty

Introduction to Use and Supply

C. A. ZRACKET
Summarized by Richard S. Greeley

Environmental changes occur continuously throughout our daily life. Major storms and floods inflict enormous human costs and tragedies. Now we are beginning to measure and document additional human costs from the increasing emissions and effluents from industrial and agricultural activity which have been presumed until recently to be associated with an increasing level of societal well-being.

As other speakers have indicated, we are becoming aware of potentially much more severe environmental changes, and we are beginning to be concerned about the potentially much larger human costs of these changes. We have learned that very great changes in the climate and other environmental conditions have occurred many times in the past (Brooks 1970). It is reasonable to expect that we will encounter similar great changes in the environment in the future. Now, in addition, human activities in search of more food and energy for an expanding world population may be leading to extensive human influences on the global as well as the regional environment. The overuse of fertilizers and the increasing use of marginal lands are specific examples of agricultural practices which may now or eventually affect climate. Thus, we may be encountering a paradox in that attempts to maintain or improve the standard of living throughout the world may result in environmental changes which harm the standard of living, at least to the extent of inflicting major human costs on certain segments of society.

There is a delicate balance between agriculture and human well-being. Any changes in climate, whether natural or manmade, will have severe consequences on agriculture and hence on human well-being (McQuigg et al. 1973). It is vital, therefore, to understand the scale of potential environmental changes, their timing, and their specific human costs in order to determine whether we can ameliorate or prevent potential tragedies.

We first have to consider that the population of the world is increasing and will probably continue to increase for the next generation, at least. World population, now over 4 billion, is projected to increase to between 5 and 6 billion by the year 2000, even if fertility decreases among the group of women now alive who will be entering child-bearing years (Weinberg and Hammond 1972), and even if famine strikes at some countries. Second, we must appreciate that industrial activity and energy use throughout the world have been increasing rapidly, at rates of 4 to 5 percent per year (at least until the recent oil price increases) (United Nations 1976; Kahn et al. 1976). The human impact on the environment undoubtedly has increased proportionately.

Therefore, we face two major global policy questions:

- Can energy and food supplies be increased and used more efficiently to provide at least the basic amenities for increasing population (or even provide an increasing well-being), or must we attempt to share a decreasing per capita supply of energy and food?
- Are there environmental limits on these attempts to increase energy and food supplies, particularly atmospheric limits?

ENERGY USE AND SUPPLY

Currently, energy use throughout the world amounts to about 8×10^3 Gwatts, a little over 0.2 Q per year.[1] This amounts to roughly one-hundredth of one percent the rate at which solar energy is absorbed by the earth's surface (8×10^7 Gwatts). The average per capita use of energy throughout the world is at a rate of about 1.7 kw. By the year 2000, with a global population of 6 billion, if the per capita energy consumption doubled (roughly one-third the current U.S. per capita use rate), the total energy use would be at an annual rate of about 2.4×10^4 Gwatts, or 0.5 Q. By the year 2100, if there were a population in the world of 20 billion each using a little more than the current average per capita rate in the United States (say, 17 kw), the total energy use rate in the world would be 3.4×10^5 Gwatts, or 10 Q per year.

Fossil energy may be able to provide as much as 200 Q, enough for 400 years at the rate postulated for the year 2000 but only for 20 years at the rate postulated for the year 2100. As noted in some of the previous sections, we may have to curtail the use of fossil energy because of the CO_2 emitted into the atmosphere. In that case, abundant energy supplies are available from solar (2,400 Q per year), nuclear fission with breeding (610 Q per 10 million tons of uranium oxide), and nuclear fusion (10,000 Q per part per million deuterium in the ocean). Realistically, by the year 2000, we could begin to use solar energy on a large scale to replace fossil fuels. Hence, to provide the 0.5 Q per year needed by the year 2000 we will, for instance, have to double oil and gas production, triple coal production, and build as many as 3,000 very large (1,000 megawatt) nuclear power plants to supply the developed part of the world. By the year 2100, for example, we could cover about 6×10^6 square kilometers with solar collectors (using 1980 technology) to pro-

[1] One Q $= 10^{18}$ BTU, a unit still in use by some writers in the United States. One BTU/year $= 3.36 \times 10^5$ watts. One Gwatt $= 10^9$ watts. The use of energy at a rate of 1 Gwatt for one year would use 8.76×10^6 kwatt-hours of energy, or roughly 30×10^{12} BTU. One BTU $=$ 252 calories $=$ 1055 joules. Energy data are from Gustavson (1975).

duce 10 Q from solar energy, assuming 50 percent efficiency of collection and 50 percent spacing of the collectors to avoid self-shading. Nuclear fission power plants with breeding and nuclear fusion power plants could also provide the required amounts of energy. Therefore, although the industrial effort and the financial resources required to build these facilities would be substantial, and the environmental impact would probably be substantial, future energy supplies need not necessarily be the limiting factor in the world's economy.

FOOD USE AND SUPPLY

Is Food a Limiting Factor to Population Growth?

Revelle (1974) has calculated that there are a potential 3.2 billion hectares [2] of arable land is the world, roughly 24 percent of the total land area. About 1.4 billion hectares currently are being farmed. He believes that an additional 1.1 billion could be brought into production. He further calculates that with multiple cropping, and allowing 10 percent of the land for nonfood products, that 4,000 to 5,000 kilocalories of food energy could be supplied per capita per day for a world population of 40 to 50 billion people. He notes that the present diet in India provides 2,150 kilocalories per day. The energy required to grow this food using agricultural methods currently in effect in India would take about one-tenth of a Q per year. Therefore, conceivably, food supplies for a very large world population may not be a limiting factor either. However, many critics have challenged Revelle's estimates as wildly optimistic, since they do not appear to have taken adequate account of the required water supply, ecological impacts, and capital constraint. (See, for example, Holdren and Ehrlich 1974, and Woodwell's discussion below.)

World Use of Fertilizers

It is recognized that increasing the use of nitrogen fertilizers can increase food-crop yields dramatically. A little over 4 million tons of nitrogen fertilizers were used in 1948 and about 40 million tons, in 1974 (Byerly 1975). Conceivably, currently arable land could produce sufficient food for 6 to 7 billion people in the year 2000 by increasing the use of nitrogen fertilizers to 100 to 125 million tons per year. Byerly (1975) noted that one bushel of winter wheat has 10 percent protein or about one pound of nitrogen per bushel. Three pounds of nitrogen applied to the soil provides one pound of nitrogen in the wheat, one pound in the straw, and one pound remaining in the ground or other waste products from the plants. In the future, it may be possible to get 2 pounds of nitrogen per bushel of wheat for 3 pounds of nitrogen applied to the

[2] One hectare = 10^4 m^2 or 2.47 acres.

soil. Therefore, until recently there was the kind of optimism reflected in Revelle's work that through the heavy use of fertilizer and improved strains of grain species, by bringing more land into production, by multiple cropping, by increased irrigation, and by augmenting land-produced crops with ocean harvests of protein from fish, and so forth, there would be no immediate limit on the world's population or well-being from a food supply standpoint.

In other words, until recently there appeared to be no shortage of total energy and food-production capability available to the world for many years of growth.

However, several questions are now being raised, particularly about food production:

- Will natural changes in climate produce severe agricultural failures?
- Will manmade effects on climate produce severe agricultural failures?
- Have we reached a point of diminishing returns in stimulating crop yields with nitrogen fertilizers and with "miracle" grains?

There are arguments that the release of the energy to support a large, affluent world population could possibly warm up the earth excessively and melt the ice sheets of Greenland and the Antarctic (Kellogg 1974; Wilcox 1975). There are other arguments that the pollution (Meadows et al. 1972) and overuse of fertilizer (Commoner 1975) to supply the large amounts of energy and food for a large population would result in global "disaster." However, such questions are still a legitimate subject for debate, and we must be cautious before accepting the predictions of either the optimists or the pessimists.

Energy, Environment, and the Economy

JOHN P. HOLDREN
Summarized by Richard S. Greeley

Figure 5 is a very simple diagram illustrating the interrelationships among energy, environment, the economy, and human well-being. Figure 6 illustrates these relationships in somewhat more detail. Energy, food, metals, and water are all needed to support the world population, and all are interrelated with the environment. Note that it takes energy to make energy resources available for use. We can expect to use more energy for this purpose as resources become depleted and hard to recover.

There are nonlinearities in each relationship as illustrated in Figure 7. A plot of benefits, such as gross national product (GNP) per capita vs. energy use per capita is not linear in any given country, and the energy needed to provide a given level of benefit varies among countries. Sweden and Norway, for instance, enjoy a high GNP per capita at a relatively lower use of energy per capita than the United States. The

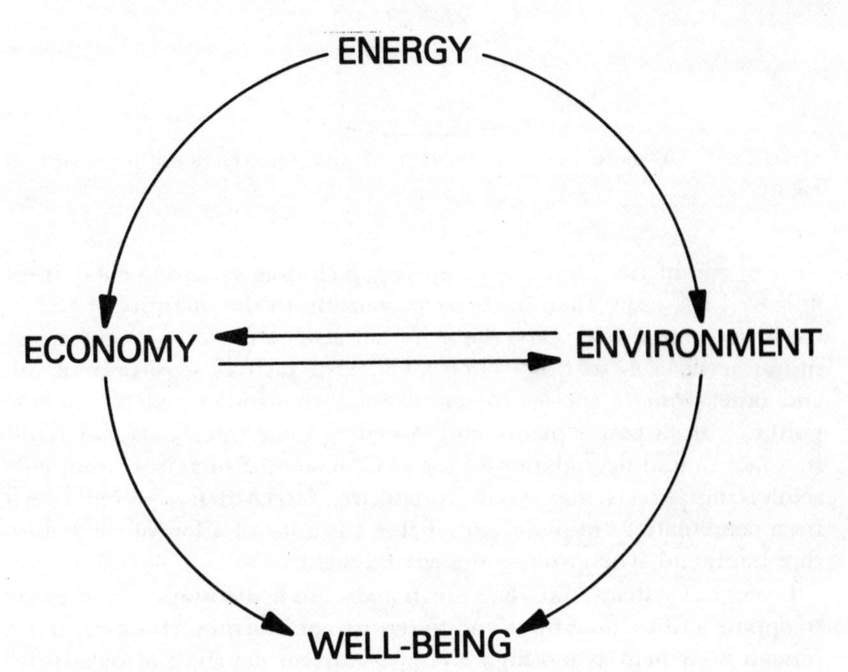

FIGURE 5. The interrelationships among energy, environment, economy, and human well-being.

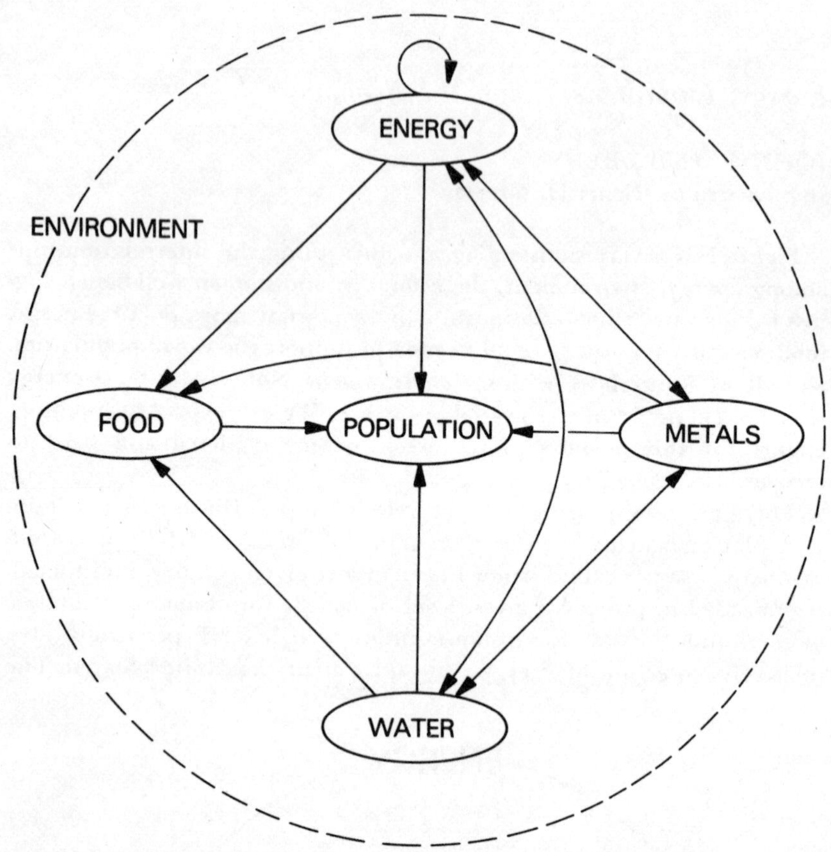

FIGURE 6. A more detailed diagram of the interrelationships shown in Figure 5.

costs of production and use of energy, including environmental costs, may well rise faster than in direct proportion to the quantity of energy used. In the past few years, the marginal economic costs of new energy supply have been soaring. Thresholds exist, such as sensitivity of fish and other aquatic species to temperature changes caused by thermal pollution from power plants, and exceeding these thresholds may result in a more rapid degradation of the environment. Synergistic (mutually reinforcing) effects may occur, producing larger-than-expected effects from combinations of pollutants. Table 1 is a list of a few of the known thresholds and synergisms in the environment.

Ecological systems, like the human body, are homeostatic, i.e., they act to oppose a disturbing force and to restore equilibrium. However, like a human body held at too high a temperature or deprived of oxygen for too long, an ecological system can collapse if sufficient stress is applied. We must ask, "How close are we to the danger point?"

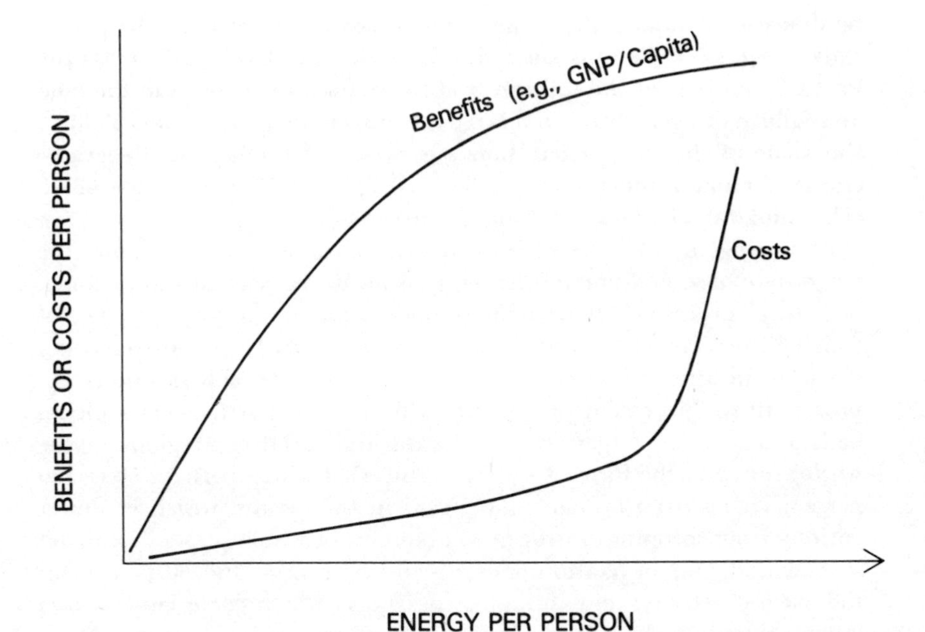

FIGURE 7. Nonlinearities in benefits and costs versus energy.

We already have reached the scale of human intervention that rivals the scale of natural processes (Holdren and Ehrlich 1974). **Furthermore, many of these forms of intervention will lead to observable adverse effects only after time lags, measured in years, decades, or even centuries. By the time the character of the damage is obvious, remedial action will**

TABLE 1. Nonlinearities in Environment's Response.

Thresholds
Temperature in Aquatic Systems
BOD in Aquatic Systems
pH in Aquatic Systems
Climatic Feedbacks (e.g., sea ice)
Synergisms
Smoking—Radon
SO_2—Particles
Acid Rain—Nitrate Runoff
Temperature—Water Pollutants
Oil—DDT

Definitions:
BOD: Biochemical Oxygen Demand.
pH: A measure of acidity of water.
SO_2: Sulfur dioxide (primarily from combustion of coal and fuel oil).

be difficult or impossible. Some kinds of adverse effects may be practically irreversible. For instance, the deforestation of tropical forests followed by erosion of the soil can lead to irreparable damage to the land and failure of agriculture in a large area for many generations. Table 2 lists some of the cause–effect time lags which can delay the observable effects of human intervention in the environment. These include physical, biological, chemical, and nuclear processes.

Returning to the scale of human interventions, Table 3 shows the magnitude of some human interventions measured against the yardstick of natural processes. Of particular concern are the following: CO_2, of which 12 percent in the atmosphere now comes from human activities; particles in the atmosphere, where anthropogenic contributions approach 10 to 20 percent of the naturally injected particles on a global basis and a much larger fraction in the industrialized regions of the world; sulfur in the form of sulfate particles, of which perhaps 50 to 100 percent comes from human sources; oil in the oceans, which is almost entirely from shipping; nitrogen compounds, of which present estimates suggest that human fixation amounts to between 10 and 50 percent of the natural activity; and chlorofluorocarbons, which come largely from human activities. Also shown in Table 3 are the trends of human intervention and the time periods during which major impacts may occur. If pollution control measures that will decrease the particulate, sulfur, oil, and mercury loading of the biosphere are instituted, then the major problems probably will come from CO_2, N_2O, and chlorofluorocarbons in the atmosphere. As noted above, CO_2 and the chlorofluorocarbons can increase global temperature through the greenhouse effect. The N_2O and the chlorofluorocarbons can react with ozone causing an increase in ultraviolet exposure on the earth. Of course, release of the chlorofluorocarbons probably can be controlled readily.

TABLE 2. Cause–Effect Time Lags.

Physical Transport
- Atmosphere
- Ground Water
- Oceans

Biological Transport
- Food Webs

Chemical/Biochemical Reactions
- Atmosphere
- Hydrosphere
- Organisms (latency: cancer, etc.)

Nuclear Reactions
- Accumulating Stocks
- Extended Dose

TABLE 3. Scale and Trends of Human Intervention.

Human Impact	Natural Yardstick	Human Input as % of Yardstick [a]	Approximate Current Rate of Increase	Trend in Growth Rate
CO_2	Atmos. pool	12	0.8 ppm/yr. 0.2%/yr.	Increasing
Particles	Input to atmos.	14	0.4%/yr.	Decreasing
Mercury	Input to bios.	10–100	Zero	Steady
Sulfur	Input to bios.	50–100	3–5%/yr.	Increasing
Oil in Oceans	Natural seeps	3–20	4–5%/yr.	Increasing
Nitrogen Fixation	Natural rate	10–50	5–10%/yr.	Increasing
Land Cultivated	Global land	11	2–3%/yr.	Steady
Desert Created	Global land	5	1–2%/yr.	Steady
Thermal Waste	Solar input	0.01	4%/yr.	Increasing
Erosion	Natural weathering	200	1–2%/yr.	Steady
Hydrocarbons	Input to atmos.	10	4%/yr.	Decreasing
Chlorocarbons (including fluorinated cpds)	Input to atmos.	>90	5–10%/yr.	Increasing
Radioactivity (Kr^{85} and T^3)	Natural amount in atmos.	0.1	6–7%/yr.	Increasing

[a] Relative to nature larger fraction in the industrialized regions of the world.

I conclude from all of this that the high rate of growth of energy use widely anticipated for the time period 1976–2000 is neither desirable nor necessary. It is not desirable because the economic and environmental costs of such growth are likely to be severe. It is not necessary because the application of technological and economic ingenuity toward the goal of more efficient energy use can produce continued and indeed growing prosperity without high energy growth (see Schipper 1976).

Impact of Environmental Change on Human Ecology

GEORGE N. WOODWELL

As human populations of the world double every 30 years, we are interested increasingly in the limitations of the resources available to support people. One of the most important resources is the fixation of carbon through photosynthesis into the earth's biota. The difference between the total amount of carbon fixed in photosynthesis, called "gross production," and the total amount of respiration by plants is "net primary production" (NPP). This is the amount of energy available from plants to support all animals, including man, and all decay organisms. The quantity of net primary production available on earth is finite and trends in its availability are of major interest to man. It is from the earth's net primary productivity that we gain, directly or indirectly, all of our food, fiber, and much of our fuel. Net primary production is one of the best criteria for appraisal of the potential utilization of biotic resources by man.

Tables 4 and 5 contain a summary of the net primary production of the major vegetation types of the earth expressed in trillion (10^{12}) kilowatt hours per year and other relevant data on average energy use. (One trillion Kwatt-hours per year equals 114 Gwatts, the unit of power used elsewhere in this report. Recall that about 8×10^7 Gwatts is the total rate of energy absorption of solar radiation at the earth's surface.)

The world total of net primary production is about 840 trillion kilowatt hours per year, of which about 500 is from terrestrial plant communities and the remainder from marine systems. Although the open ocean contributes a large fraction of the total from marine systems, the net productivity *per unit area* is very low. The net production obtained from plant communities of the coastal zone is very much higher, frequently approximating the 1,000–2,000 grams of dry organic matter per square meter per year available from forests.

A careful analysis of the extent to which the earth's net primary production is being used directly in support of man leads to the conclusion that, at present, as much as 50 percent of the net production is being used in support of human food supplies. Considering first the resources of the oceans, present indications are that the catch from the world fisheries has leveled off at about 70 million tons annually. While the issue can be argued that sound management and improved technology may increase that yield, the yield appears to have reached a maximum at present. This means that we are harvesting from the oceans all of the net primary productivity that is available to us. We have no other ways

TABLE 4. Net Primary Production of the Earth. Net Production is the Amount of Energy or Organic Matter Available from Plants to Support Animals (including man) and Organisms of Decay. (Adapted from Whittaker and Likens 1973).

	10^9 t C./yr	10^{12} kwh/yr
Continental		
Tropical Rainforest	15.3	162
Tropical Seasonal Forest	5.1	55
Temperate Evergreen Forest	2.9	36
Temperate Deciduous Forest	3.8	45
Boreal Forest	4.3	53
Woodlands and Shrublands	2.2	27
Savanna	4.7	49
Temperate Grasslands	2.0	21
Tundra	0.5	5.8
Deserts Scrub	0.6	7.0
Rock, Ice and Sand	0.04	0.3
Cultivated Land	4.1	43
Swamp and Marsh	2.2	23
Lake and Stream	0.6	7.0
Total	48.3	534
Marine		
Open Ocean	18.9	237
Upwelling	0.1	1.2
Continental Shelf	4.3	50
Algal Beds	0.5	5.8
Estuaries	1.1	12.8
Total	24.9	307
World Total	73.2	841

TABLE 5. The Flux and Average Density of Energy Worldwide.

Source of Energy	Worldwide Flux kwh/yr	Density kwh/m²/yr
Solar Energy		
Top of Atmosphere	156×10^{16}	3.05×10^{3}
Net Production of Plants		
World	841×10^{12}	1.4
Temperate Forests and Agriculture	—	5–10
Nonbiotic Energy		0.088 Worldwide
World (1967)[a]	44.9×10^{12}	0.301 Land only
U.S. (1967)[a]	15.6×10^{12}	1.67
Manhattan	—	$\sim2.4\times10^{3}$ [b]
Kings (Brooklyn)	—	$\sim1.36\times10^{3}$ [b]

[a] Man's Impact on the Global Environment (SCEP), MIT Press, 1970, p. 294.
[b] Estimated on basis of per capita use of energy in U.S.

at present of using carbon fixed in the oceans other than by harvesting fish populations that use it as food.

An analysis of the uses to which man puts terrestrial vegetations reveals that most of the grasslands of the world have either been transformed into agriculture or are grazed now and used directly in support of people through food production. Forests are harvested for fiber, and there is now virtually no forest that has not been made available for this purpose. We also use firewood, probably more on a worldwide basis than we use energy from fossil fuels. This superficial analysis of the extent to which man is now using the earth's net primary productivity justifies our statement that direct uses probably exceed 50 percent of the world total.

Indirect uses of the world's biota are more difficult to define. They include the stabilization of water flows, the stabilization of water quality, and the stabilization of the soil surface. The vegetation affects certain qualities of air as well as water and controls in some degree the temperature and moisture regime of entire regions. These functions in the stabilization of essential environmental resources are often overlooked. When they are lost, society either suffers a loss in the quality of environment or is forced to make an accommodation in the form of dams to control water flow in rivers or filtration plants to correct for losses in water quality.

The fact that the toxic effects of human activities are spreading worldwide and reducing the structure of the biota is an indication that human activities at present exceed the capacity of the biosphere for repairing itself. One of the best examples is the increasing acidity of rain in certain segments of northeastern North America and in Scandinavia. While it is difficult to measure direct effects of this increasing acidity on the indigenous vegetation, the inference drawn from experience with soils and with the effects of toxins on plants leads experienced ecologists to assume that long continued acid rains will lead to serious losses in the capacity of forests and agriculture to fix carbon in photosynthesis. A 10 percent loss in net primary production is difficult to measure, even in agriculture. A 10 percent loss over the area of the six New England states would be a loss of as much fixed energy as that produced by 15, 1,000 megawatt nuclear power plants. The loss would appear in reduced crop production, in reduced fisheries, in reduced lumber production, in a decrease in the water retention capacity of forests, an increase in river and lake acidity, and in other ways. The loss would represent a diminished capacity of the earth to support people.

One way to appraise whether a loss in net primary production is occurring worldwide may be through analysis of carbon dioxide data. Figure 8 is a graph of the monthly CO_2 concentration in the atmosphere

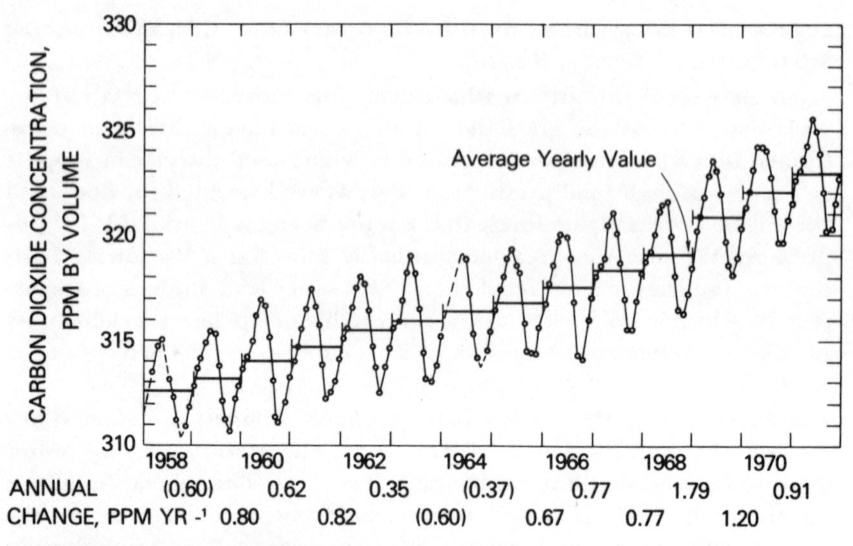

FIGURE 8. Mean monthly carbon dioxide concentrations at Mauna Loa. Annual changes in parentheses are based on incomplete records (Keeling et al. 1973). Reprinted with permission of *Tellus*.

above the trade wind inversion at Mauna Loa, Hawaii, taken by C. D. Keeling over a period of 15 years (SMIC 1971; Machta and Telegadas 1974). The concentrations of CO_2 in air vary seasonally; they reach a peak in late April and a minimum in September or October. The difference in amplitude between winter and summer is about 5 ppm at Mauna Loa, due apparently to the removal of CO_2 from the atmosphere by photosynthesis during the northern summer. Data for the Southern Hemisphere are similar but are 6 months out of phase and have a smaller amplitude, apparently due to the smaller land area in the Southern Hemisphere. This observation, coupled with recognition that about five-eighths of the net primary production of the earth occurs in terrestrial ecosystems, suggests that the winter–summer oscillation in CO_2 content of air is due principally to the storage of carbon in terrestrial ecosystems. This observation raises the possibility that a decrease in the net primary productivity of terrestrial vegetations would appear as a reduction in the difference between the peak concentration of CO_2 in April and the low concentration of CO_2 in the fall over a period as long as a decade. Measurements derived from the data available do not show such a change, despite the strong presumptive evidence that such a change must have occurred. Additional techniques and more refined analyses are required in the exploration of this important problem.

The major challenge is to establish a stable relationship between the resources used in support of human populations and the demands placed on those resources. Can we develop man-dominated systems that

do not degrade the biotic systems on which they depend? To establish such stable relationships the man-dominated systems must have interactions with the rest of the biosphere that are similar to the interactions that the natural ecosystems of that place had prior to their destruction. Under such circumstances, the biotic matrix of the earth within which man lives and on which he continues to depend can be maintained indefinitely. Without it, the problems of management of air and water and land and the biota itself become increasingly difficult and ultimately impossible.

Climatic Variability and Its Impact on Food Production

STEPHEN H. SCHNEIDER
Summarized by Richard S. Greeley

The major attribute of climate for food production is its variability. Figure 9 is an illustration of a thousand-year history of Iceland's mean annual temperature, deduced from ice records of the harbor of Reykjavik. Temperatures there have varied by roughly 1 °C during this period. Figure 10 shows five scales of estimated temperature data for a variety of regions and latitudes. Paris-London temperatures, Greenland ice cores, White Mountain tree rings, and central England measurements all show significant departures from "normal." Average temperatures deduced from instrumental observations in the Northern Hemisphere indicate an increase of about half a degree Celsius from 1880 to 1940 and then a decrease of about one-quarter of a degree until the present.

In the United States we recently have enjoyed a long period of relatively good weather for agriculture. For the last 10 to 15 years, weather in the Great Plains breadbasket has been less variable in summer than during most of the recorded history in the United States. The Midwest drought in the 1950s lasted roughly 2 to 5 years. The dust bowl of the

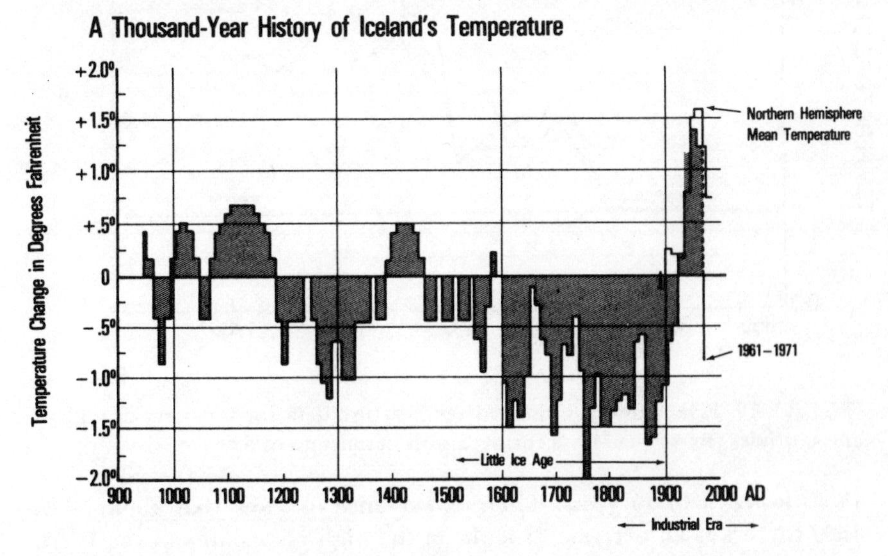

FIGURE 9. A 1,000-year history of Iceland's temperature. (D. Gilman, National Weather Service. Presented at 140th meeting of the American Association for the Advancement of Science, San Francisco, 27 February 1974.) Reprinted with permission of AAAS. (After R. A. Bryson and P. Bergthorson.)

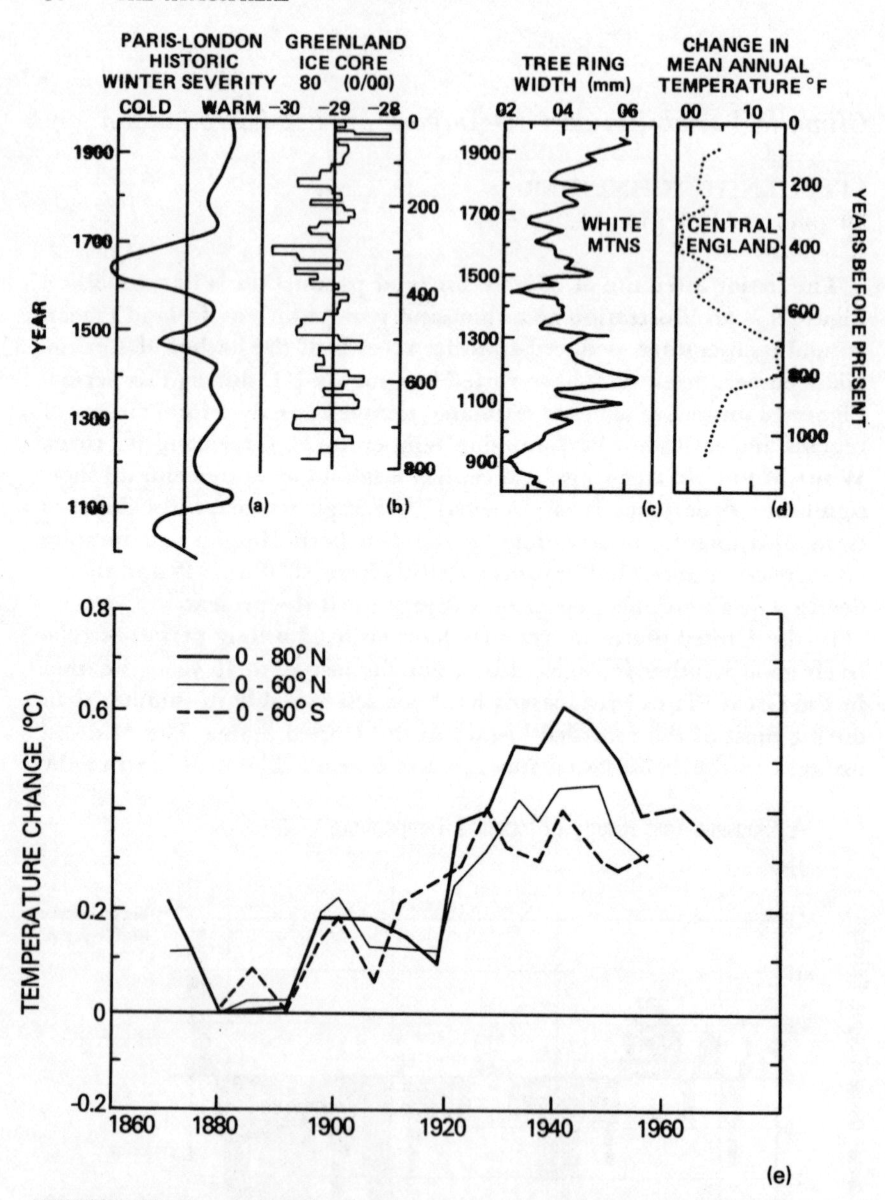

FIGURE 10. Five scales of estimated temperature data for a variety of regions and latitudes (Bryson 1974) . Reprinted with permission of *Science*.

1930s lasted 5 to 10 years. There is evidence to show that about A.D. 1560 there was an extreme drought in the high plains area of the West, and deposits of windblown dust reached 3 meters in Western Nebraska during that period (Weakly 1962).

The recent good weather has resulted in relatively good food production in the United States. Since 1955, as shown in Figure 11, for Missouri

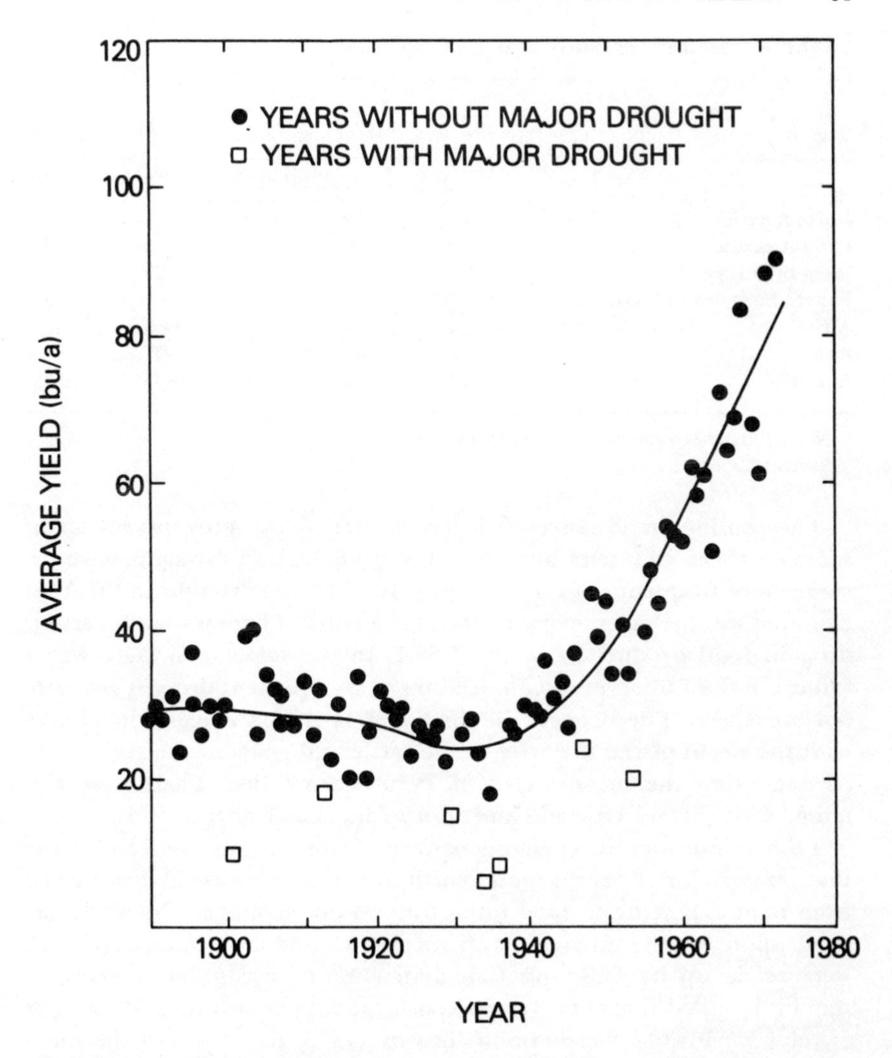

FIGURE 11. Average Missouri corn yields (*The American Biology Teacher.* 1974. 36:534–540) . Reprinted with permission of *Am. Biol. Teach.*

corn yields, Midwestern agriculture has shown very high yields. The summer rain has been above average and summer temperatures have been below average, both favorable in this granary.

Abrupt changes in climate and abrupt changes in food production have historical precedent. For example, around 550 B.C. the Northern European winter abruptly worsened, leading within about 3 years to a shortening of the growing season by 20 to 30 days. The shorter season persisted for centuries (Brooks 1970; Roberts 1975). The "Little Ice-Age" between about 1550 and 1850 was a period of recurrent crop failures in Europe and England.

TABLE 6. The Changing Pattern of World Grain Trade.

Region	1934–38	1948–52	1960	1970	1976 (prel.)
	(million metric tons)				
North America	+5	+23	+39	+56	+94
Latin America	+9	+1	0	+4	−3
Western Europe	−24	−22	−25	−30	−17
Eastern Europe and USSR	+5	—	0	+1	−25
Africa	+1	0	−2	−5	−10
Asia	+2	−6	−17	−37	−47
Australia	+3	+3	+6	+12	+8

Note: Plus=net exports. Minus=net imports.
Source: Brown 1975.

The cooling trend since 1940 has shortened the growing season in some northern countries by as much as 2 weeks, and droughts have become more frequent than in the 1960s. In 1972, the drought in the Moscow area was the most severe in about 300 years. There was an 8 percent drop in food production in the USSR. In the same year there was a failure in the Indian monsoon, leading to an 8 percent drop in rice production there. The drought in the Sahel of Africa reached its climax with the death of the majority of the cattle and goats in the region. At the same time the anchovy crop off Peru nearly failed. There were also minor droughts in Australia and South America (Roberts 1974).

Table 6 indicates the changing pattern of world grain trade and shows that, except for Australia and North America, the world has moved from food exporting to food importing. Many people in the world depend upon North American crop surpluses. The crop failures of 1972 were made up by U.S. and Canadian exports and grain reserves. It should be noted that total grain exports amount to less than 10 percent of total worldwide consumption. Brown (1975) has reported the number of days that reserve stocks could feed the world. This number has dropped from 105 days of reserves in 1961 to 31 days estimated for 1976. Note that there are essentially no areas in the United States previously idled for grain production which now could be brought easily into production. Thus, it is vital to determine how vulnerable is North American agriculture to climate variability—the controlling factor in crop yields.

Figure 11 illustrates the remarkable increase in average corn yields in Missouri since 1940. Of particular note is the difference between the years with and without drought. Recently, yields have all been above average, possibly due to the better-than-average weather.

Figure 12 shows one of the major reasons for the increasing yields—namely, the increasing amounts of fertilizer used in the United States.

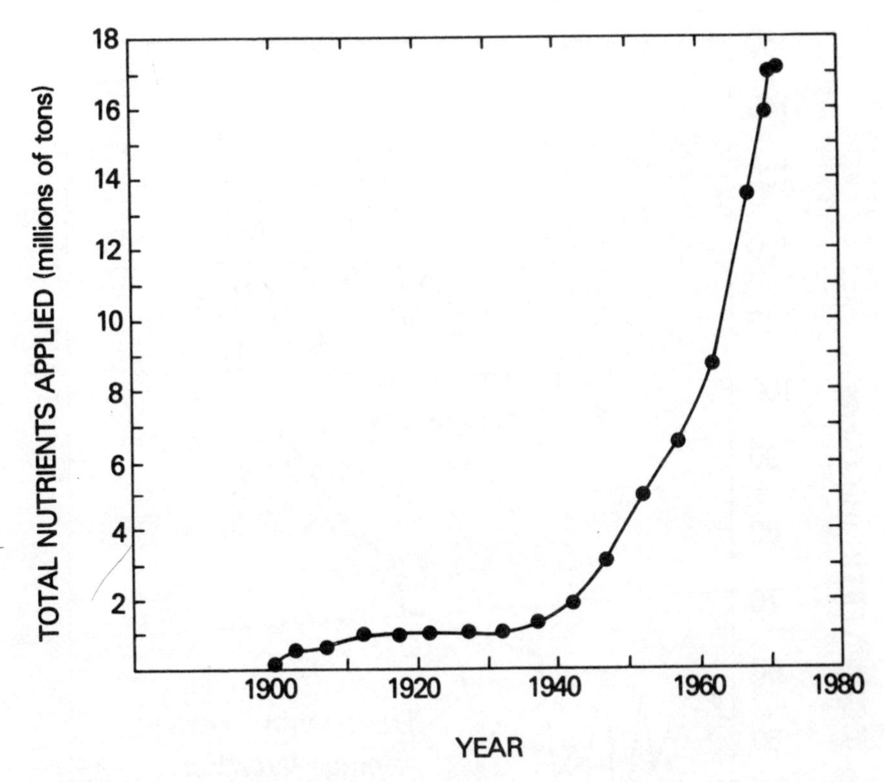

FIGURE 12. Trend of fertilizer use in the United States *(The American Biology Teacher*. 1974. 36:534–540). Reprinted with permission of *Am. Biol. Teach.*

Another reason is the use of improved plant strains. Have we begun to reach the point of diminishing returns in the use of fertilizer and development of improved species? Figure 13 indicates that average corn yields in Iowa are beginning to approach the yields of experimental plots. Commoner (1975) warned that overuse of fertilizer is extremely harmful to the soil and the watersheds into which the excess nitrates run off. He says there may be dangers to people, particularly children, who eat food such as spinach containing excess nitrate. Under certain bacterial conditions the nitrate can be converted to nitrite which can cause poisoning. High nitrate levels in drinking water are also dangerous to human health. Byerly (1975) disputed Commoner's views but recognized that a problem exists.

Another important point is the increasing amount of energy used in the United States to grow crops and bring food to the table. Figure 14 is a plot of farm output versus energy input. A leveling off since the late 1960s would indicate that using a great deal more energy in farming will probably not increase yields appreciably—without new technologi-

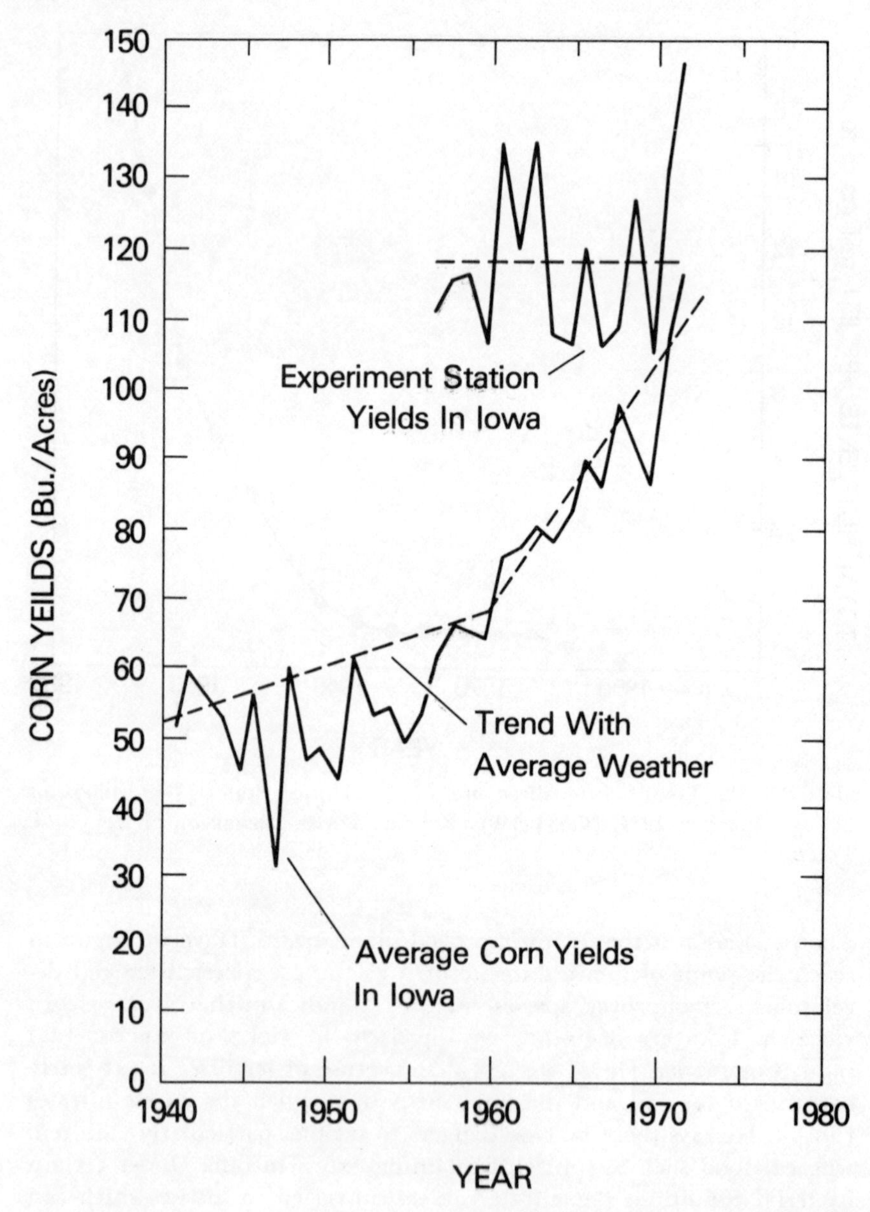

FIGURE 13. Average corn yield in Iowa in relation to yields of experimental plots (L. M. Thompson. 1975. *Science* 188:535). Reprinted with permission of *Science*.

cal investigations. Figure 15 illustrates that the United States increasingly has "subsidized" farming with external energy sources since the early 1900s. The ratio is now roughly 10 calories supplied to the farming system for each calorie in the food on the plate. Figure 16 shows a

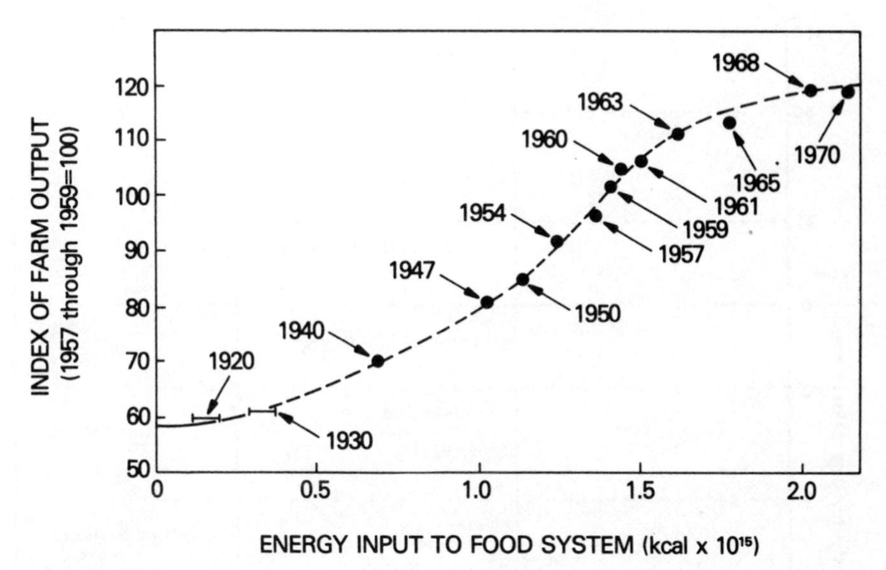

FIGURE 14. Farm output as a function of energy input to the U.S. food system, 1920–70.

comparison of the energy input to calorie output among different types of food production methods. The most efficient is wet-rice culture, which gives 50 calories out for each calorie in—500 times more efficient than feedlot beef, widely used in the U.S. diet.

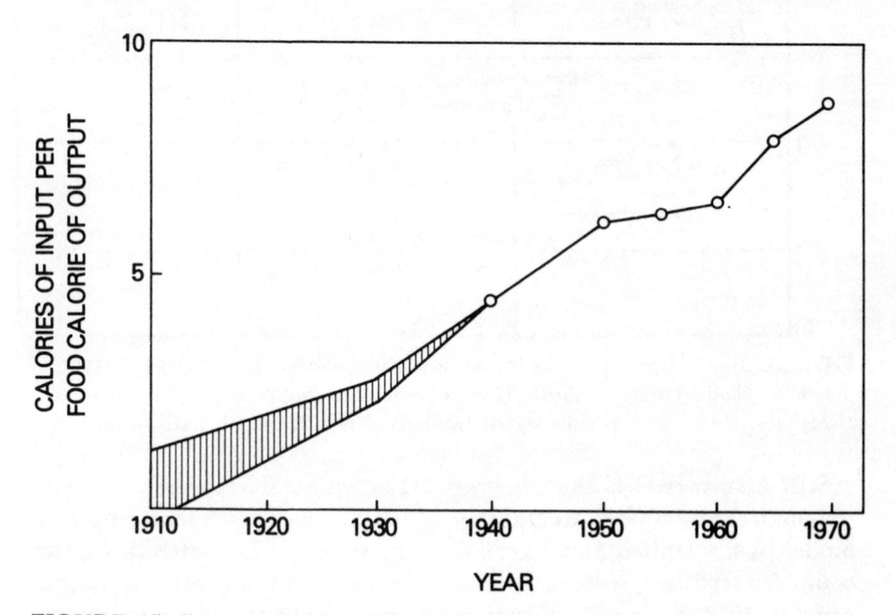

FIGURE 15. Energy subsidy to the food system needed to obtain one food calorie.

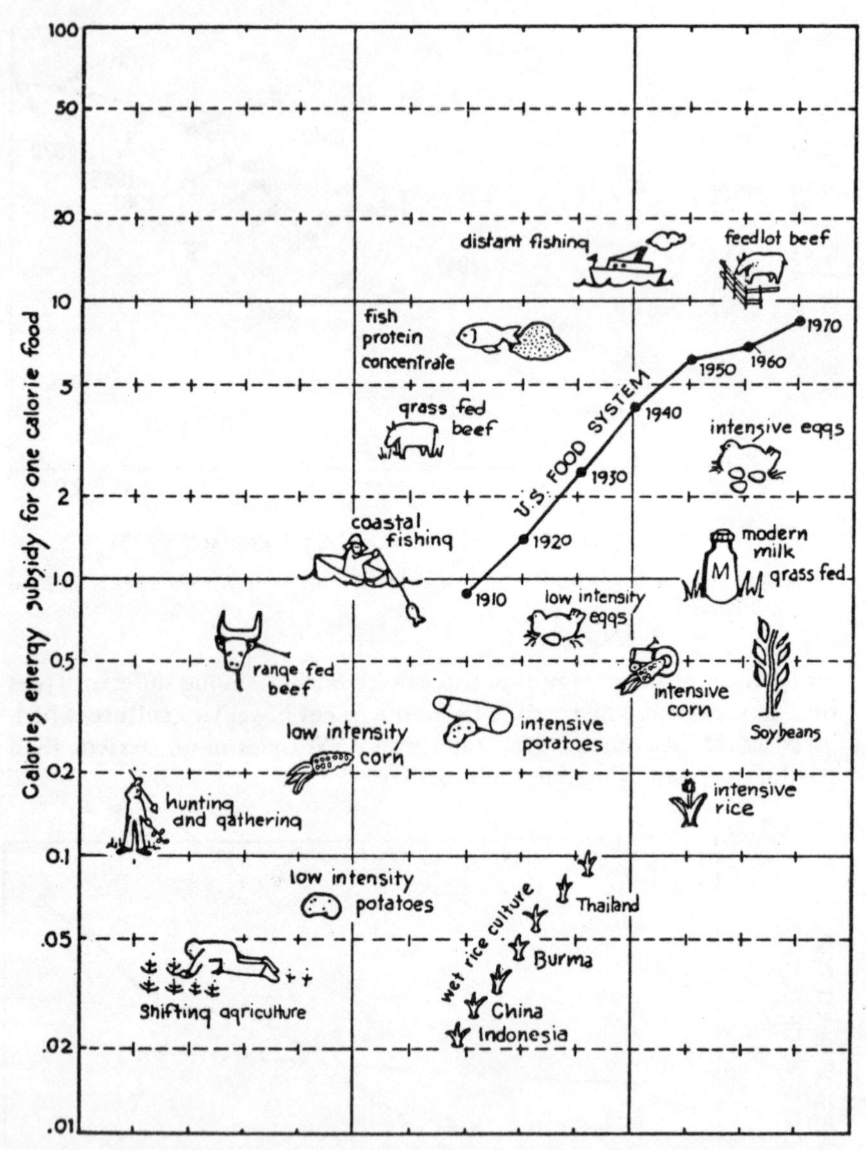

FIGURE 16. Energy subsidies for various food crops. The energy history of the U.S. food system is shown for comparison (Steinhardt and Steinhardt 1974). Reprinted with permission of Duxberry Press, Belmont, California.

Still unanswered is the question "How vulnerable is the U.S. farm system to the weather?" McQuigg et al. (1973) developed a semi-empirical model of the United States corn farming system, which attempts to account for fertilizer, species, and other human technological innovations prior to 1973. Figure 17 illustrates the results. The plot shows average yields of corn per year since 1900 "corrected" to 1973 technology. The

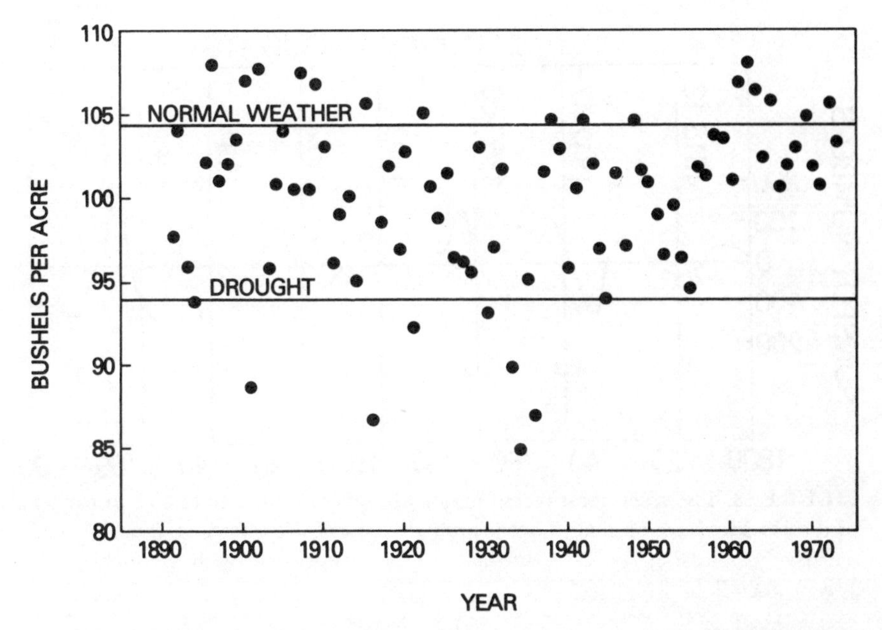

FIGURE 17. Simulated five-state, weighted average corn yields using 1973 technology and harvested acreage: Ohio, Indiana, Illinois, Iowa, and Missouri.

only remaining variable is the weather. The "normal" value is about 105 bushels per acre. However, unusual conditions can give yields as low as 85 bushels per acre or even lower.

Of particular note is the unprecedented high yields since 1958 due to the unprecedented good weather over that 15-year period. McQuigg et al. (1973) estimated that the probability of having another period of 15 years of such good weather to be 10,000 to 1 against. Sure enough, 1974 and 1975 were not particularly good weather years. Record total production in 1975 was achieved primarily by large increases in acreage brought under production, not by improvements in yields per acre, which dropped substantially in some cases.

Figure 18, which plots sunspot numbers vs. year since 1800, also shows a remarkably regular succession of droughts in the plains states east of the Rockies. It is tempting to believe that there is some physical explanation for this correlation, though no satisfactory cause-and-effect mechanism has been discovered so far. Without stretching the correlation too far, it would seem that we are "due" for another drought soon, whether or not we accept the sunspot connection.

In theory, the so-called "miracle" strains of crops can be either more sensitive to severe environmental conditions—in Fig. 19, or less sensitive—in Fig. 19. In either case, a degradation in environmental conditions will mean a greater variability in *total* food production than traditional crop varieties, even if *percentage* yield-variability decreases.

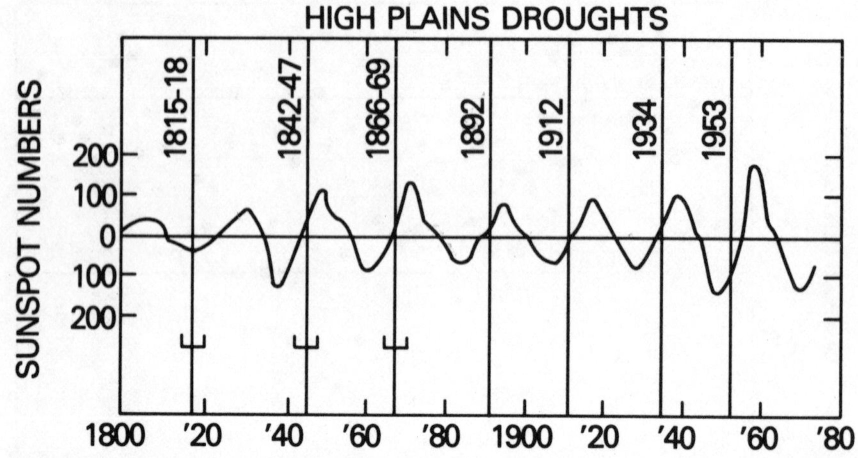

FIGURE 18. The seven most recent major droughts in a five-state area in the lee of the Rockies have tended to occur after the end of the alternate (here plotted negatively) sunspot cycles (Marshall 1972). Reprinted with permission of J. R. Marshall.

Thus, the probability of a large drop in food production in the United States should be of real concern. One cannot look just at the recent past to predict the future. We must look further back to obtain realistic estimates of the variability in the weather in order to make better actuarial calculations of food production fluctuations.

IS THERE A FOOD-CLIMATE CRISIS?

The current margin between food supply and food requirements throughout the world is uncomfortably small, as was shown in the previous section. Figure 20 illustrates the food and population balance in the developing countries. Per capita food production is barely staying even. Table 7 is a list of the millions of people in regions of the world with insufficient protein supply.

Severe food shortages are almost inevitable in the next decade or two, given any reasonable hypothesis of climate and weather variability. Polar cooling, whose progress up to 1972 is shown in Figure 10, may correlate with greater frequency of extremes of drought, flooding, and other anomalies, and in dislocations of the large-scale circulation that could have particularly serious implications for the semi-arid and otherwise marginal lands into which husbandry and settlement are increasingly migrating (Bryson, 1974). This is a controversial point, however. The USSR will find it very difficult to be independent agiculturally in the long term if their food consumption habits continue to rise. The USSR is also far more vulnerable to climate change than the United States because of its location at a higher latitude. The USSR is con-

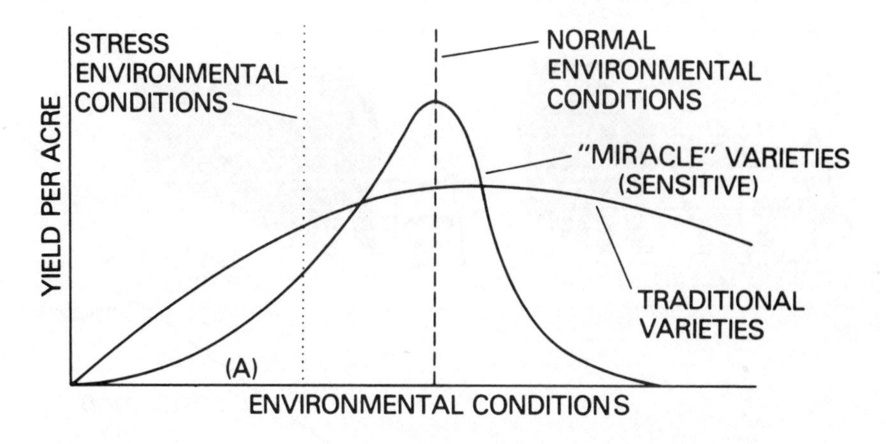

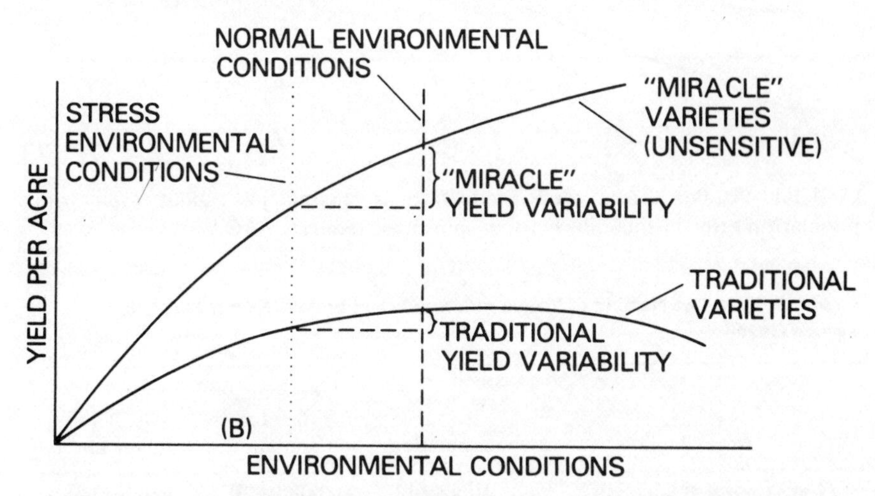

FIGURE 19. So-called "miracle" strains of crops can be either more sensitive to severe environmental conditions as in (A) or less sensitive, as in (B).

sidered to be a food importer in nearly every imaginable scenario of the coming 25 years (Roberts 1975).

The United States and Canada are slated for increasing roles as the world's food suppliers. Yet this country has no visible contingency plans for not-so-improbable severe weather changes and crop failures. We appear to have no explicit, long-term policies regarding use of additional fertilizer, increasing cultivated acreage, irrigation supplies, or developing newer species except more of the same. We appear to have no explicit policies regarding food reserves, price supports, or stabilizing food prices. Research on climate change and the reduction of variability of

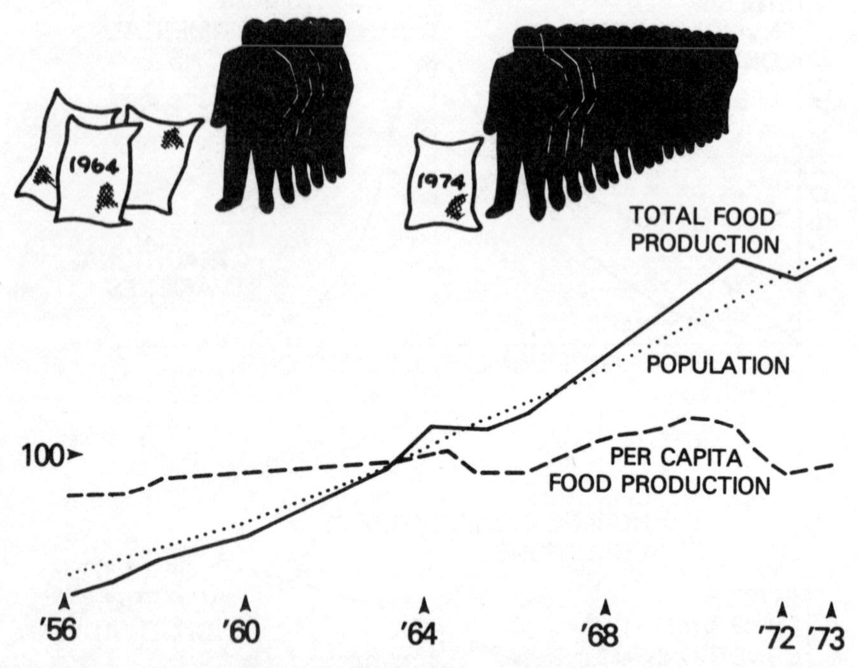

FIGURE 20. Per capita food production is staying just about even with population growth in the developing countries. (Source: FAO.)

TABLE 7. Estimated Number of People with Insufficient Protein/Energy Supply by Regions (1970).

Region	Population	Percentage Below Lower Limit	No. Below Lower Limit
	(thousand million)	(percent)	(millions)
Developed Regions	1.07	3	28
Developing Regions Excluding Asian Centrally Planned Economies	1.75	25	434
Latin America	0.28	13	36
Far East	1.02	30	301
Near East	0.17	18	30
Africa	0.28	25	67
World (Excluding Asian Centrally Planned Economies)	2.83	16	462

Source: FAO

A high percent of the populations of the poorer nations, excepting the Republic of China, suffer inadequate nutrition.

food production is not being forcefully advanced. (See, for example, Schneider 1976.)

There is a further fear that mankind's industrial and energy production activities may affect the climate and lead to enhanced probabilities of extreme variability. Thus the food-climate crisis could be very near-term and of major significance. It is also a longer-term problem, in that the currently available solution such as bringing additional land into crop production and using much larger amounts of nitrogen fertilizer can also influence the climate, and they will have limits not yet foreseen in total, long-term productivity. The smallest impact, and one we have already seen, is the triggering of higher prices for food by crop failures in one nation, such as the USSR in 1972, which had to be made up by North America. Figure 21 is a plot of food prices from 1965 to 1973. Simultaneous crop failures in North America and the USSR could lead to even higher prices and widespread starvation throughout the world. Figures 22 and 23 illustrate some of the consequences. Some estimates

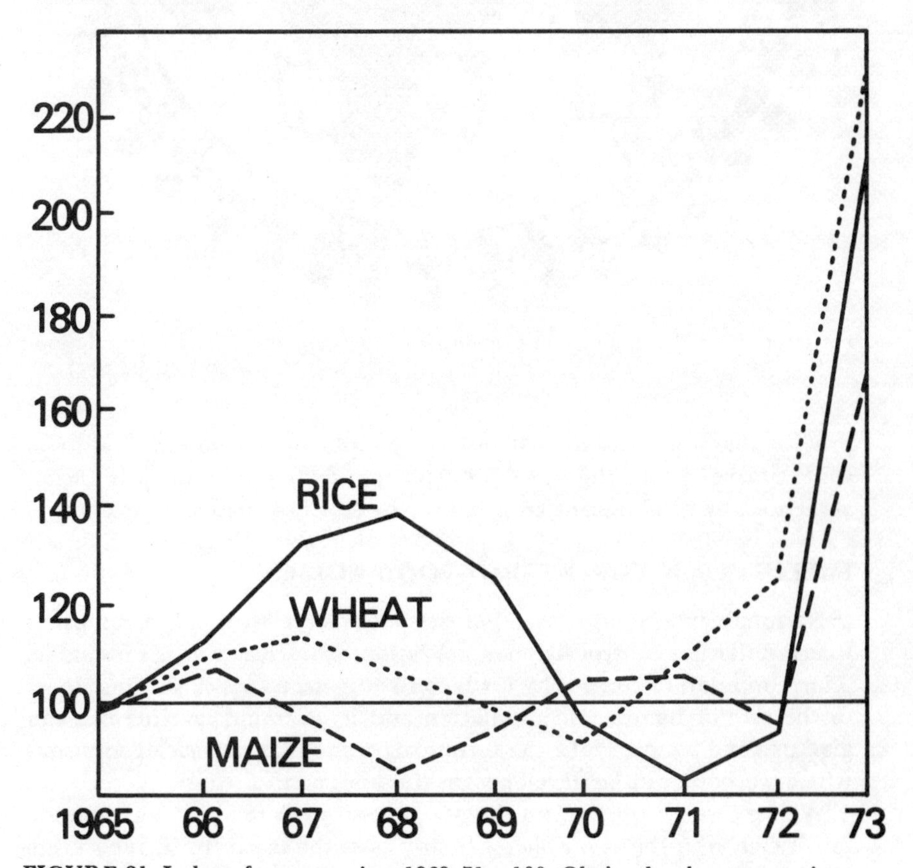

FIGURE 21. Index of export prices 1969–71 = 100. Obviously, the poor nations suffered the greatest human impact. (Source: FAO.)

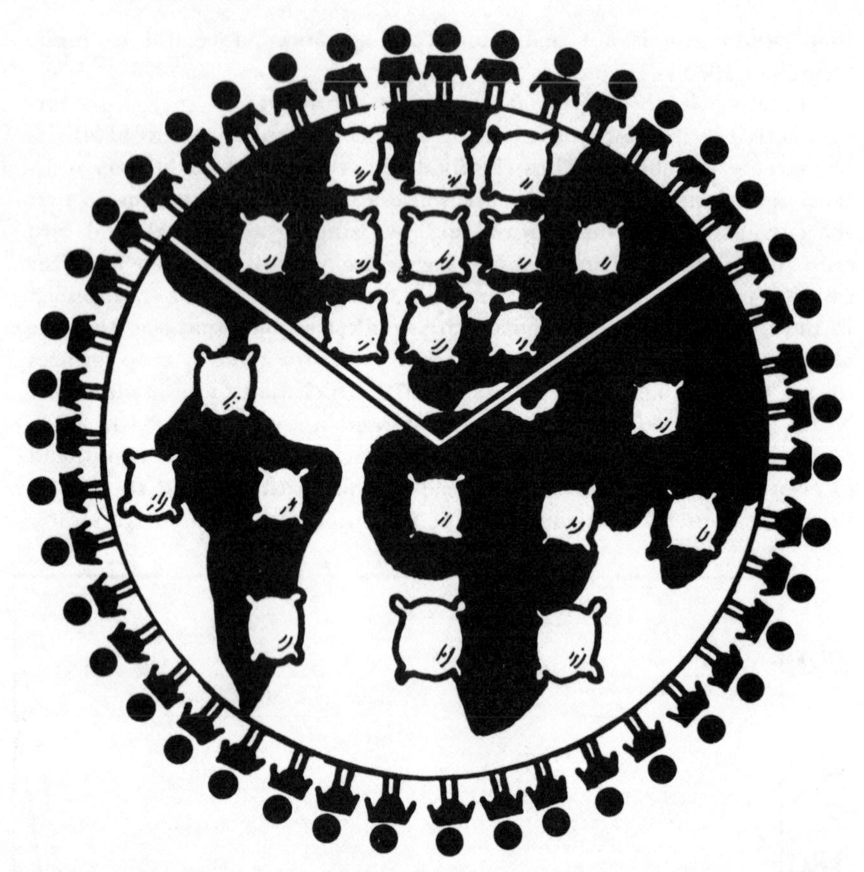

FIGURE 22. The high-income countries (39% of the world's population) accounted for 51% of the total consumption of cereals for all uses in 1969–71.

predict that upwards of 100 million people in developing countries could starve, while the more affluent countries would be just inconvenienced by a significant crop failure in North America.

IMPLICATION FOR ENERGY-FOOD POLICY

National energy and food policies must start with the assumption that population control by mass starvation or nuclear war is untenable. Thus, population control by birth control is necessary. A slowing down in the rate of increase of population and its eventual leveling out can give us time to introduce the technical, economic, and social solutions which we hope can be developed in the long run.

We may well want to limit the use of fossil fuels to avoid emitting too much CO_2 into the atmosphere. In any case, the use of fossil fuels eventually will be self-limiting simply by their availability. Nuclear fuels may prove to be environmentally acceptable at a certain level of risk.

FIGURE 23. It takes more grain to produce adequate nutrition via the corn-fed beef cycle than it does by direct human consumption.

In any case, solar energy needs to be brought into use wherever possible.

We need to learn very soon whether large increases in the use of artificial nitrogen fertilizers must be curtailed, and if so, what alternatives exist for increasing yields of cereal grain crops. Alternatives may have to be devised for the present system of wealthy countries using a high proportion of meat in the diet, raised from cereal grains in a very inefficient overall process (see Fig. 23), while poor countries are left with the bare minimum of cereal grains for direct use. In other words, meat substitutes made from such items as soy beans, avoiding the very inefficient process of raising beef cattle, may be necessary to even out the use of protein between rich and poor populations. Such alternatives may require major social changes for their implementation (Holdren and Ehrlich 1974; Schneider 1976).

Part 4

Summary of the First Day's Discussion

J. DANA THOMPSON, *Rapporteur*

Following the contributions of the speakers on the first day of the Conference, participants discussed what they were to do and accomplish in the remainder of the 3-day meeting. The discussion evolved naturally from a period of questions and comments concerning Dr. Lovelock's Gaia hypothesis (summarized in Appendix II). Rather than attempt to summarize that discussion topic-by-topic I have chosen to present it as it developed, in chronological order. I have included all relevant comments—as completely and in as much detail as possible. No attempt is made to identify each person who participated in the discussion. I consider this discussion to be of much value to an understanding of what the Conference was about and what it accomplished.

A participant asked if someone were going to tell us about the effect of a change in the ozone layer. Why are we afraid of it? What will it do to living creatures?

Dr. Lovelock replied that it is alleged variously that the ultraviolet radiation that would reach the surface if there were no ozone layer could do anything ranging from causing discomfort to destroying all life on earth. He cited a growing body of evidence suggesting that most life on earth is quite tolerant to ultraviolet radiation. Certainly there are very large variations over the surface of the earth. In a mountain region near the equator there are very high levels of ultraviolet exposure—but it is doubtful whether one would see a sunburned tree there. Nevertheless, one does not, with callous calm, accept willingly the modification of a significant sector of the atmosphere without wondering what is going to happen.

Another scientist pointed out that the National Academy of Sciences (NAS) studied this problem and prepared a rather disturbing report which says that there probably would be an increase in the incidence of skin cancer roughly proportional to the flux of ultraviolet erythemal radiation at the surface, and that the effect would be more serious among light-skinned people than dark-skinned people. More troubling, the report cites possible disruption in the ecosystem due to any *differential* susceptibility of different kinds of organisms to ultraviolet light. In an ecological system it is not necessary to disrupt everything to cause a serious effect. Rather, a change in species composition can propagate through the food web to influence that level important to human beings. One reason to believe these ultraviolet changes might not be innocuous is that both DNA and proteins have strong absorption bands in the ultraviolet erythemal spectrum. It is rather hard to imagine, said a participant, that under those circumstances, where molecules common to living cells are particularly adept at absorbing energy, serious biological effects will result.

Dr. Lovelock strongly criticized the NAS report, and noted that the changes in the ultraviolet radiation one obtains by moving around are extraordinarily large and there is no evidence of the sorts of changes described in the NAS document. One should not forget that ultraviolet light is different from other dangerous radiations, such as X-ray or nuclear radiation, and is definitely beneficial in low levels. To speak of ultraviolet radiation as analogous to nuclear radiation is most misleading, he said.

The fundamental question arose of how society undertakes policy and decisionmaking in the face of scientific uncertainty. In the case of the ozone question are scientists to decide when the evidence justifies a call for policy action, knowing that worthwhile studies that could be made in the next 10 years may not shed much more light than at present? Should a decision be made based on the incomplete theoretical evidence which indicates that large changes in the ozone layer can be expected in 10 to 20 years if we do nothing?

Another suggestion was that we recognize that some issues cannot be studied completely in a time that is short compared to the time it takes for a potential effect to be felt by society; therefore, one issue the Conferences should address is how scientists can aid policymakers in the face of such uncertainty. One participant objected because he did not accept this as an appropriate issue for the Conference.

Dr. Mead called for a "cease fire" in an attempt to avoid a premature polarization of the participants. She asserted that the Conference was organized to consider the policy implications of findings about the atmosphere and the environment generally. Can scientists state anything in any form that will be of use to policymakers? Clearly, it is going to take longer to study some of the problems we are discussing and make the answers satisfactory *to scientists* than it is going to take to make policy decisions concerning these problems. The time interval required before we begin to see clear evidence of a particular manmade effect on the environment may be long compared to the time in which society has to act.

One scientist said that he understood the task of the Conference to be primarily to consider what *kinds* of statements could be made about the dangers to the atmosphere, and the methodology scientists should use in order to be sure that those statements are responsible and acceptable to the public. He contrasted that with the task of identifying what the present dangers to the atmosphere *are*.

By way of clarification, Dr. Mead said that the question is, "can scientists agree on a position that would be useful to policymakers?" Are there any points where the reflection of endless academic arguments could be useful in decisionmaking? Instead of insisting that we need 100 years of experimentation, is there anything we can agree on about margins of danger concerning environmental hazards? Are there even minimal statements we can make about the uncertainty and risks of proceeding in a certain way?

Dr. Mead emphasized that the Conference was based on the assumption that policy decisions of tremendous depth are going to be made—*whether scientists provide input or not.* There is no way for scientists to avoid affecting the decisionmaking process on issues related to their discipline, even if they remain publicly silent. A decision by policymakers *not* to act in the absence of scientific information or expertise is itself a policy decision, and for scientists there is no possibility for inaction, except to stop being scientists.

With that clarification, the conferees addressed more specific questions. One observed that levels of probability and levels of risks were important factors in decisionmaking. It may be essential to know some things with a very high probability of certainty because an error in judgment could be disastrous. On the other hand, lower-risk decisions do not require such certainty. For example, a spacecraft requires an extremely high probability of good performance of its life-support systems, but such high probability of performance is not required of a minor device such as a camera. Therefore, in a sense, deciding what you want to know becomes a policy decision and is a critical link between scientist and policymaker.

One participant wondered whether the Conference was organized with the preconceived notion that environmental change was automatically dangerous and bad. Do we equate change with danger—or are we looking for the good that might come of change?

The first response was that, based on Dr. Lovelock's paper, change in either direction might have deleterious consequences. It was pointed out that changes caused by mankind all occur on short time scales and they occur in a world that has evolved biologically on a very long time scale. Nature's repair processes occur on the long time scale, too. There is an advantage in stabilizing the relationships between basic resources and our day-to-day needs. The unparalleled increase in human population and its demands for food, energy, and resources is clearly the most important destabilizing influence in the biosphere.

Regarding the desirability of change, a participant suggested that if we know absolutely nothing about the effects of change, then we might assume that 50 percent will be bad and 50 percent will be good. The question then becomes: What is the potential magnitude of any change for which we want to worry about the 50 percent that will be bad? The next question is: What is the magnitude of the damage that it might cause and is it irreversible? Finally, we ask: Do we have any reserve capacity to recover from or bypass the damage? For example, consider the case of potential ecological hazards that may affect the food chain for man. Our world food situation is so strained already that one might conclude that any 50 percent risk of bad effects is a serious danger.

Those comments were attacked as misleading, like saying that "when I stick my pencil into my watch and stir it around there is a 50-50 chance I will improve it." When we are dealing with a biological system which is rather finely tuned in many respects and which has evolved over a long period of time, the odds are much higher that a given perturbation will cause a negative effect. The importance of time scales cannot be emphasized enough. Some people tend to argue that "evolution is the solution to pollution." Yet when you look closely at how evolution tends to solve things you find it solves them with extreme mortalities per generation. Mosquitoes, for example, evolve resistance to DDT—yet the mortality rate for several generations is enormous. If you are talking about people, that is precisely the sort of event we are interested in avoiding.

These comments were reiterated by another participant who also observed that public arguments about why man is affecting nature often are based on spurious evidence which stirs up emotive reactions in the public and is sensationally reported by the press corps. Scientists often are accused of crying "wolf" and perhaps the general public is beginning to think that we are not to be believed, that we tend to fly off the handle. By all means, let us say that disturbing the ozone layer by even 1 percent is bad—and leave it at that. Unless we have good evidence, we do not need to explain all the reasons why it is bad.

A member of the press commented that if scientists say publicly that something imperils life but do not say why, they cannot expect to be listened to or be taken seriously. The problem is not that some of the disasters scientists have predicted have not come about; the problem is that the signal tends to get lost in the noise. For example, consider that the breast cancer rate has been going up about one percent per year. Clearly, there are reasons why this is happening—it is not necessarily the pill or a diet rich in animal fats—it is probably a combination of many factors. Nevertheless, it is useful for scientists to say candidly, "We don't know why such-and-such is happening, but it is happening and here are some of our suspicions why." It is not useful to claim that reporters are irresponsible—they have to have *something* to say about what is going on.

One scientist asked the reporter if it is "really reasonable to publicly present the fact that breast cancer is increasing and then spew out a litany of spurious assumptions or assertions as to why this might be so?" The reporter replied, "You must use the best knowledge that you have at the time and say this is the best possible guess we have."

Another participant suggested that if the consequences are unknown, but the system is so delicate that it is a mistake to monkey with it, that is a reason for concern. We have enough data to know that the ecosystem is fragile. Saying the system should not be monkeyed with until you know more is giving an answer, but it is not an answer in terms of, say, skin cancer, which is problematic.

It was pointed out by another scientist that the Climatic Impact Assessment Program, instituted by the Department of Transportation to look into the potential effects of supersonic aircraft, must report its findings to Congress in 3 years. This is a political deadline and not a scientific one. Scientists who participated in the study were concerned that information released prematurely, about which they were not 95 percent certain, might be interpreted incorrectly. Assume, they said, that you are driving a car filled with people down a desolate desert road with very little gas and one fork in the road leads to a gas station and the other to a dead end and serious trouble. You ask the passengers, "Do I go left or do I go right?" Only one person says, " I was here 20 years ago—I don't remember very well—go left." You go left. In a sense we have an analogy here. Why are we afraid to set out the evidence? If our evidence is very sketchy, let us say as much. Let us try to defend it only up to the point that it deserves defending. The scientist continued, "I get very upset at the idea that we should make the unilateral decision as scientists to withhold some piece of evidence, which may have some policy effect, because we, in our personal philosophies, do not think that evidence is sufficient to justify action. Let others make that decision. Put out the evidence but make sure you state its uncertainty."

Another participant commented that if we frequently make uncertain statements to the public, we are going to be wrong more frequently—and as a group we may be trusted less. This is a serious dilemma, because if we do not make statements, we make policy by default—probably worse policy than would be made otherwise. If we are going to be useful to society, we must find ways of allocating resources for science with greater promise of utility in the process.

In response, it was reiterated that we must indicate how reliable the information is. If a group of scientists says, "we don't think this information is very reliable, but the net result would be very bad if it is true," it is difficult to imagine that the public would be tremendously upset.

Another reporter discussed the dilemma of deciding which scientists to believe on an issue. In the chlorocarbon issue, for instance, does the press corps believe the scientists who say we are doing harm to the ozone layer, or the DuPont Company scientists who say there is no harm being done? It is the obligation of the media to report both sides as objectively and fairly as possible until the issue is resolved.

"Are we going to examine the consequences of not monkeying with the environment?" was the next question put to the Conference. Should we look at the problem from the point of view that there are going to be at least twice as many people in the world in the next century as there are now? How are these people to be fed and clothed? If we do not monkey with the ecosystem, how do we feed these people? Dr. Kellogg replied that this was an important point for the Conference to consider and noted that the question of doing nothing *or* doing something both involved risks and difficult policy decisions.

This turned the discussion perceptibly, and several people questioned whether the ecosystem really is as fragile as we think. One person noted that we have been screaming this at the public for so long that we now have an obligation to be more objective about "the delicacy of the ecosystem." We are finding that polluted lakes can rejuvenate at remarkable speed and that most pollutants added to the atmosphere are removed or rendered harmless within a relatively short time. An ecologist countered that Dr. Broecker had given an excellent example of how the biosphere is *not* compensating for the activities of man (see Appendix I). Man-produced CO_2 is not being taken up by the biosphere at a rate comparable to its production rate. Further, the evidence that biotic systems are sensitive to human activities is overwhelming. No one is ever going to repair the damage done to the fisheries of the Great Lakes and most of the rivers of the East Coast of the United States. No one will ever repair the damage to our eastern forests—the loss of the chestnut tree was a very serious loss, both economically and ecologically. There are many other examples. Man is having important, lasting effects on the ecosystem, he said.

A conferee noted that man always has affected his environment. Unfortunately, since the emergence of fire and agriculture, he has increased his influence. Today, the problem is the speed with which man is altering the ecosystem, and there may be some rate of alteration beyond which the ecosystem is subject to collapse. This group should stress that the ecosystem has never been called upon to react to changes that occur at today's rate.

In conjunction with those comments, one scientist wondered if the scale of alteration to the ecosystem required to feed an ever-increasing population was not so large as to overshadow many of the popular environmental issues being discussed. For example, is worrying about a one percent change in ozone in the stratosphere "straining at gnats and swallowing camels?"

But no one at the Conference actually proposed that *nothing* be done in the face of obvious environmental alteration by man. Rather, it was pointed out that we recognize there are large risks in doing some of the things we have been doing. Although we are constrained by the population and its growth rate to provide food and resources, there are enormous differences in how we go about it. The issue is not "to try or not to try;" the first issue is what set of criteria or constraints should we apply to the technologies we focus on those needs? Are we obliged to take into account as a main criterion for evaluating those technologies their impact on the environment? Are we to minimize the threats to what we believe is a delicate ecosystem? Second, to what extent do we start influencing the derivatives—the rate of change and the rate of change of the rate of change? If we listen only to the technological optimists who tell us how easy it is, in theory, to feed 10, 20, or even 30 billion people, then no attempt will be made to deal with those derivatives. We will end up in a sense "biologically committed" to a preposterous attempt to reach those population sizes which surely will lead to an environmental catastrophe.

Following those comments, a participant asserted that the United States is presently exporting nitrate and phosphate at a rate we cannot hope to continue. Exported grains contain this nitrate and phosphate, which we must then replace with fertilizer. It was noted that we cannot continue indefinitely to use our land the way we have been using it, and we may be faced eventually with the prospect of just feeding our own people.

A second comment concerning man's potential environmental impact was that calculations at Lawrence Livermore Radiation Laboratories showed that man had the capability, through full-scale nuclear war, to destroy from 30 to 70 percent of the ozone layer due to the added oxides of nitrogen. About 10 years would be required for nature to repair the ozonosphere following such a nuclear war.

The session was concluded with the thought that we as a species are trying to maintain ourselves at the expense of other species; there seems to be a conflict between preserving nature and feeding the rapidly increasing population. Is our major objective really to feed the population, or do we realize we cannot continue to feed the world *at any price?* Where do we strike a balance between preserving nature and feeding the world?

Part 5

Managing the Atmospheric Resource: Will Mankind Behave Rationally?

BARBARA WEST, *Rapporteur*

A summary of material presented by:

Harrison S. Brown
Edith Brown Weiss
Wendell A. Mordy
Jack E. Richardson
Margaret Mead
William W. Kellogg
Stephen H. Schneider

The point made again and again in the previous chapters is that, increasingly, mankind's activities, whether intentionally or not, are modifying the atmosphere on which all depend. There are no forces or customs barriers that can be thrown up to keep another nation's air or water pollution out or a desired quantity of stratospheric ozone in; increasingly, nations can protect only "their" atmosphere by protecting *the* atmosphere, which is shared by all. This is a kind of problem with which our previous notions of individual ownership or national sovereignty have not equipped us to deal (see Margaret Mead's remarks in the Preface). Let us now look at society's uses of the atmosphere, the history of regulation and international cooperation on the atmosphere, national and international concerns, and what mechanisms might be effective to begin to handle these problems.

SOCIAL USES OF THE ATMOSPHERE

Individuals "use" the atmosphere in essentially passive ways, breathing it and adapting their activities to the patterns of water, sunlight, and temperature which surround them.

Societies, on the other hand, are considerably more active users. The physical properties of the atmosphere are utilized for radio communications. Aircraft fly through it. It is a dump for wastes ranging from the CO_2 we breathe out to industrial gases and particulates such as chlorofluoromethanes, DDT and PCB, krypton-85, and ash particles. It is used as an essentially "free" resource, a source of water and one component of the recurring patterns we call climate. The military uses the atmosphere not only for all of the above purposes but also for other activities such as reconnaissance and, on occasion, delivery of a variety of weapons and chemicals bringing death and destruction. It is the entry and departure point to and from outer space, not only for satellites and space probes but also for energy from the sun and elsewhere in space.

Finally, and increasingly in the last few decades, we have been exploring ways to modify the weather and also discovering that we have been modifying it as an unintentional outcome of other activities. Over 60 countries have experimented with modifying the weather, and perhaps a quarter of these either have or are considering an operational program in some aspect of weather modification. In the area of inadvertent climate modification, cities are consistently warmer than the surrounding countryside and, in mid-latitudes of the Northern Hemisphere, the areas generally east of large cities get less sun and more rain than those upwind to the west. As we have been learning in the last few years, our continuing use of chlorofluoromethanes can be expected to have a significant effect on the ozone layer, thus on the amount of ultraviolet radiation reaching the earth. That the precise effects are not yet known does not guarantee that they will be trivial (see part 2).

Increasing awareness of the extent and impact of inadvertent weather modification, plus the new technology of weather and climate modification, will raise important political problems which will demand new responses from the international community (Kellogg and Schneider 1974).

HISTORY OF THE REGULATION OF USES OF THE ATMOSPHERE AND COOPERATION IN EXCHANGE OF METEOROLOGICAL INFORMATION

A "law of the land" has developed over the millennia and now is extensively codified within concepts of ownership, sovereignty, and national interests. (Antarctica and the moon are two, and the only two, significant exceptions in this body of law.) A Law of the Sea has evolved over hundreds of years, and includes concepts of both national and international rights. The ongoing U.N. Law of the Sea negotiations are grappling with questions of changes in these patterns to reflect either increased perceptions of ownership or of communality in the utilization of this shared resource.

The "law of the atmosphere" is, in comparison, very primitive, but some international law for specific areas has developed. In the communications field, the frequency spectrum has been divided up, basically on a first come, first served basis, and the International Telecommunication Union (a specialized agency of the United Nations) has grown up to oversee this use pattern. In the area of transportation, the Paris Air Navigation Agreement of 1919 and the Chicago Air Convention of 1944 firmly established the concept of national air space over which a state has exclusive sovereignty. Succeeding international agreements have strengthened and refined this concept.

In contrast to these two areas, where rights of "ownership" of uses of the atmosphere are clearly agreed upon, international law on the use of the atmosphere as a dump for noxious wastes, and the related obligations and liabilities of one state to another for compensation for adverse effects, is in its infancy.

The first, and one of the few, examples of international atmospheric problem resolutions occurred in 1941, when the United States and Canada set up a joint tribunal to adjudicate the case of a smelter, in Trail, British Columbia, emitting sulfur oxides that caused damage to crops in the United States. The tribunal issued a cease-and-desist order to the owners of the smelter, awarded damages to the aggrieved farmers, and ordered continuing monitoring of the air quality.

On a regional basis, one significant, ongoing program has been undertaken. Norway, Sweden, and Denmark have committed themselves to the

Nordic Air Pollution Convention. This provides that they will notify one another of the intent to build any structure which might have polluting activities and allows the citizens of one country to go into the courts of another to seek redress for damages. For the instances of air pollution arising within their own jurisdictions, this Convention is working well. It does not, of course, provide either relief or protection from pollution generated by other countries, such as the acid rains experienced in Norway caused by pollutants from Western and Eastern Europe.

To date, successful international adjudication of atmospheric issues has tended to be regional—and generally among nations with shared borders—leaving some very large issues unaddressed and unresolved. Encouragingly, global cooperation has been achieved in the exchange of current meteorological data. The history of international cooperation in meteorology deserves a close look as an instructive example of how international cooperation can evolve and be extremely effective in its self-defined task.

Formal international cooperation in meteorology dates to the First International Meteorological Congress in Brussels in August 1853, which was called in response to the growing need to standardize meteorological observations from ships for maritime transport. The conference adopted a set of standard instructions for making the necessary observations and a standard form for recording them. A suggestion to standardize land observations was regarded as premature. Twenty years later, increasing interest in meteorological research, greater recognition of the economic importance of climatic data, and the development of the electric telegraph, which facilitated rapid collection and dissemination of observations, resulted in the standardization of meteorological observations made from land. The Leipzig Conference of Meteorologists in 1872, which was responsible for standardizing the land observations, also established a permanent body to handle meteorological problems common to the international community.

By 1873, when the First International Meteorological Congress was held, governments had begun to view meteorology as a science which must involve cooperation. Representation at the conference no longer was by meteorologists in their private capacity, but was limited to government representatives. The second International Meteorology Congress in 1879 set up an international committee of nine nongovernmental experts and established a pattern of voluntary cooperation between meteorologists on international problems. From then until 1950, international cooperation in meteorology largely was a nongovernmental effort conducted on a voluntary basis.

Before World War I, advances in international cooperation in meteorology came in research and in the establishment of technical commissions to standardize the gathering and cxchange of data. The first instance of international cooperation in meteorological research was the International Polar Year from 1882 to 1883. Twenty states conducted studies primarily on three topics, one of which was tropospheric meteorology. During this time technical commissions were firmly established to coordinate and standardize meteorological observations and the exchange of data.

After World War I, the development of radio and of aviation greatly facilitated the gathering and transmission of meteorological data and consequently made governments increasingly aware of the economic significance of meteorological forecasts. This led to a shift in the pattern of international meteorological cooperation from one involving only nongovernmental experts to one expressly involving governments. The Conference of Directors of the International Meteorological Organization (IMO) decided in 1935 that, in the future, meetings of the International Meteorological Organization would involve government representatives and requested governments to designate directors of national meteorological services to represent them at these meetings. After World War II, when meteorology became significantly more visible as an international concern, the International Meteorological Organization was transformed into an intergovernmental organization (the World Meteorological Organization), which it remains today.

After World War II, new technology gave rise to expanded international cooperation in meteorology. The primary concern was to issue revised codes for transmitting weather information worldwide. The expansion of commercial airline services to include jets created new demands for meteorological data from the high levels of the atmosphere for use in aviation weather forecasting. This in turn led to further international cooperation in making high-level meteorological observations. From July 1, 1957, to December 31, 1958, the International Geophysical Year was held to conduct studies of the earth and upper atmosphere. The research resulted in the first daily weather maps of the world, the first comprehensive look at Antarctic weather, and the discovery of three countercurrents in the ocean.

Technological developments after World War II also brought about a reorganization in the relationship between the many organizations which had at least a peripheral interest in meteorology. The IMO was faced with establishing new agreements to replace the ad hoc relations that previously were maintained in order to avoid the increasing duplication and overlap.

The United Nations General Assembly considered the subject of meteorology and of weather modification for the first time in 1961. In September 1961, President John Kennedy proposed before the United Nations General Assembly "further cooperative efforts between all nations in weather prediction and eventually in weather control." The Assembly adopted a resolution on the peaceful uses of outer space which listed two main purposes of efforts in atmospheric sciences: (1) more knowledge of the basic physical forces affecting climate and of the possibility of large-scale weather modification; and (2) the development of existing forecasting capabilities to assist states to make more effective use of their capabilities through regional meteorological centers. During the 1960s, international cooperative efforts in meteorology increased. In 1960, President Eisenhower had invited the Soviet Union and 20 other countries to make supplementary observations of cloud cover from the Tiros II satellite when it passed over their territory. In August 1961, an international meteorological satellite workshop was held to increase the ability of countries to use the satellite data. A special U.S. task force was set up to look for possible paths of cooperation in meteorology with the Soviet Union.

After considerable negotiations by scientists, the Soviet Union and the United States in 1962 signed a bilateral accord, the Dryden-Blagonravov Agreement. It called for both countries to coordinate the launching of their operational weather satellites for maximal coverage of the weather and to establish a new communication link between weather centers in Moscow and Washington. The agreement stated in part: "In the field of meteorology, it is important that the two satellite launching nations contribute their capabilities toward the establishment of a global weather system for the benefit of other nations." For several years only conventional data were exchanged. In August 1966 the Soviet Union began to exchange satellite cloud pictures and infrared data; this exchange was interrupted for several months but was resumed in March 1967, and is still in progress. Launchings, however, are not yet coordinated.

At the request of the United Nations, the World Meteorological Organization and the International Council of Scientific Unions drew up plans to extend meteorological forecasting and cooperative research in meteorology. Two programs emerged: The World Weather Watch (WWW) and the Global Atmospheric Research Program (GARP). They are some of the finest examples of international cooperation which exist today. The main catalysts for the World Weather Watch and the Global Atmospheric Research Program were the development of satellites and of high-speed computers for transmitting data. Both of these came at a time when there were increasing demands for meteorological data for use in economic and social development. The World Weather Watch provides for the observation, processing, and transmission of meteorological data between countries in order to give better weather forecasts. States transmit data to the World Weather Watch in prescribed codes. Of about 135 countries, all but five supply weather information in compliance with the World Weather Watch procedures. Reportedly, countries make available approximately 95 percent of all the information that they should make available. The Global Atmospheric Research Program is a large-scale research effort to understand the global weather system in the hope of increasing the length of weather forecasts and eventually of understanding large-scale weather modification. The Global Atlantic Tropical Experiment, part of GARP, took place in 1974. A global observational experiment originally was scheduled for 1977, and now is to take place in 1978–79.

While progress in international cooperation in meteorology has been impressive and remarkably steady, it is important to note that some measures requiring even more international cooperation were consistently rejected until the last two decades. As early as 1873, there were unsuccessful proposals to establish an international institute to collect, process, and publish data from observation stations across the globe and to establish an international fund which would allow placing observation stations in remote places of the globe. These proposals were rejected repeatedly as requiring more international cooperation than was either possible or desirable. Not until the 1960s did revolutionary technological developments and increasing appreciation of the importance of meteorology provide a hospitable terrain for the introduction of the World Weather Watch and the Global Atmospheric Research Program.

The history of meteorology suggests an emerging recognition by states of their common interest in the global climate and weather systems. Professor E. K. Federov, Chief of the USSR Hydrometeorological Service, explicitly recognized the global community interest in weather and climate in a lecture he delivered before the WMO's Fifth Congress in 1967:

> It is not difficult to understand that the problem of transforming the climate on a world or regional base scale is, by its very nature, an international one, requiring the united efforts and the coordination of the activities of all countries. Ever more rapidly humanity is approaching the stage in its symbiosis with nature, when it can turn to practical account all the natural resources of the earth and when, as a result, it will become capable of thinking in terms of natural phenomena on a planetary scale . . . it is hardly necessary to prove that, in these circumstances, all mankind should regard itself as a single whole in relation to the surrounding world. There is no other way.[1]

Some recent events indicate that portions of the international community finally are moving toward recognition of pollution and atmospheric modification as significant global issues. The U.N. Conference on the Human Environment, held in Stockholm in 1972, included in the final documents statements of two principles and one recommendation relevant to this discussion. Principle 21 calls on states to ensure that activities within their jurisdiction or control do not cause damage to the environment of other states or areas beyond the limits of their national jurisdiction. Principle 22 calls on states to cooperate to develop international law regarding liability and compensation for victims of pollution and other environmental damage.

Recommendation 70 from the same U.N. Conference calls on governments to be mindful of activities in which there is appreciable risk of effects on climate. To this end it calls for states to:

- Evaluate carefully the likelihood and magnitude of climatic effects from a contemplated action, and to disseminate these findings to the maximum extent feasible before embarking on such activities.
- Consult fully with other interested states when activities carrying a risk of such effects are being contemplated or implemented.

Two years later, in September 1974, the USSR proposed a U.N. resolution dealing with the stratosphere and harm to the ozone layer. In response, the U.S. Department of State issued a draft treaty in August 1975 which called on states "to prohibit hostile acts of war against the environment."

---

[1] This material, with slight modification, is from Edith Brown Weiss 1975. International responses to weather modification. International Organization 29 (3), Univ. of Wisconsin Press, Madison.

While this recital of the development of international concern and cooperation includes some promising steps and indicates a rising level of awareness by states of the potential for international harm from national activities, it must be noted that so far states have not bound themselves to any enforceable, multilateral treaty which effectively limits their actions in the areas of weather or climate modification, or which commits them to indemnification to other states for actions resulting in intentional or unintentional modification of the atmosphere.

It is instructive to note the outcome of the one test of the Stockholm resolutions in this area. Australia and New Zealand, upset at France's continuing atomic weapon testing in the atmosphere, went to the World Court to seek relief. They cited a number of international accords to which France is a signatory, including Principle 21 of the Stockholm Convention, and asked the Court to prohibit further testing by France in the Pacific. After due deliberation, the Court agreed that France should not have conducted its atmospheric testing. And there they stopped—no prohibitions, no sanctions. Shortly thereafter, France detonated another device.

The history of international cooperation suggests that before states will seriously consider cooperative arrangements they must reach a critical threshold of foreseeing gain from such efforts which they cannot obtain with their own resources, or of seeing the threat of a "decreasing sum game." This means that scientific development must be sufficiently advanced that states can foresee either benefits from cooperative arrangements to develop the field further or seriously adverse consequences from employing the technology without arrangements for coordination and mutual restraint.

NATIONAL VERSUS INTERNATIONAL CONCERNS

What are the areas in which states currently view actions as well within their sovereign powers, yet the results of those actions reach across national borders?

The industrialized West is the primary user and producer of both chlorofluoromethanes and fertilizers, both of which catalyze the reaction which destroys ozone (see part 2). The results of such ozone destruction would affect the entire earth, yet we consider it our right to continue the manufacture and use of these substances without consulting other nations. (Perhaps ironically, the incidence of skin cancer is likely to be highest in the West simply because susceptibility is partly a function of skin color. No such "just desserts" effect is known to exist for plants, however.)

The USSR has vast areas of potentially productive land where average annual yields are low because of a lack of adequate water. They also have large quantities of fresh water, virtually unused, flowing north into the Arctic Ocean. And they need, or at least strongly desire, to produce more food. What could be more reasonable than to divert these rivers to the semiarid regions to the south? Indeed, they are planning to do just this. But what will happen to the world's weather patterns, so many of which are spawned in the Arctic, when the flow of fresh water into that ocean is significantly decreased? Models of the earth's climate suggest that it may have a noticeable warming effect, since it could decrease the amount of sea ice in the Arctic Ocean—though this is still hypothetical. Suppose that in the first or the fifth summer after the rivers are diverted, China or the Midwest has a severe drought or unprecedented floods. Was this caused by, aggravated by, or independent of the change in the quantity of fresh water flowing into the Arctic Ocean?

In spite of an inability to prove it conclusively (at least with the present state of the art), meteorological conditions which have been attributed to nature or to God will now be blamed on man and his institutions. People will become skeptical that weather patterns are natural and not the results of intentional modifications. There may be numerous charges of attempts to modify weather and the effects of these attempts. Any intervention in the weather system may become a political act. This could only increase the incidence of conflict between states and among different parties.

Given that human activities are now influencing the atmosphere in significant ways across national boundaries, what are reasonable goals for managing these influences? Broadly, they might be defined as preserving and, where possible, enhancing the value of the atmospheric resource to all users. This, to be effective, will involve the development of a much broader appreciation of the possible international effects of actions which heretofore have been regarded as well within national rights. In the long run a new sense of mutual responsibility by each state to the others is called for, with the atmospheric system regarded as a global resource.

To implement this approach, a series of policies will have to be developed, accepted, and made operational. These policies might be grouped under the following six categories:

Information: **Free and open gathering and exchange of data in order to understand and predict changes in the atmosphere and in climate.**

Assessments: **Including warnings of impending atmospheric, weather, or climate disasters.**

Procedures and processes: **Consultation between states on forecasted or proposed changes, including the likelihood of impact of inadvertent weather and climate modification.**

Responsibility: **Acceptance by states of responsibility for changes they make in atmospheric patterns, whether intentional or inadvertent. This will require processes for consultation, monitoring, and the resolution of disputes.**

International scientific assessment for large-scale projects which could affect climate: Including a ban on large-scale field experiments without the informed consent of affected states.

Processes for conducting weather-modification experiments: Including notification, consultation, monitoring, assessment, and responsibility of the initiating countries for adverse effects inflicted on others. This also will carry the responsibility for the richer countries to aid those without adequate scientific resources in evaluating not only the direct effects of their activities but also their implications in a variety of fields (such as changes in trade patterns).

How, or whether, nations come to be convinced of the need or desirability of mutual concern for the shared resource of the atmosphere remains a question. Several efforts and organizations already exist which handle some aspects of some of the problems we have noted; other problems are quite untouched by any international concern.

Some problems in the shared use of the atmosphere, as noted earlier, are most amenable to *regional* action. (See also appendix V.) The United States and Canada can and have arrived at equitable solutions to common problems. For other countries to be involved in, for example, the establishment of air quality standards along the U.S.–Canadian border would seem to add no advantages and certainly could complicate the process.

As the Nordic Air Pollution Convention moves to grapple with pollution from other countries, it probably will attempt to deal directly with offending nations rather than encumber the process with full-scale international machinery. For example, significant amounts of pollution are being deposited in Scandinavia by "acid rain" originating in Great Britain and West Germany. Expanding the scope of regional atmospheric cooperation to include more of Western Europe is potentially a viable means to address this problem. Thus, it is possible to visualize levels of regional organization dealing with appropriate atmospheric issues—generally attempting to keep the organizational structure commensurate with the problem's scope.

To the extent that a problem can be identified as purely regional in impact, the countries involved should be encouraged to handle it expeditiously on that basis. Many regions will need the information-gathering and processing facilities of other countries or organizations to carry out such programs; World Meteorological Organization (WMO) with its established links to nearly all countries would appear to be an ideal international vehicle for such a task.

Other problems are apparently regional, and are being approached on that basis, but the effects may extend beyond the cooperating states. The Philippines, People's Republic of China, Japan, and the United States currently are considering a weather modification program aimed at mitigating the effects of typhoons. Each year the Philippines in particular faces, and frequently receives, severe damage from these storms. China and Japan receive a significant proportion of their water budget from these same storm systems each year. If it is possible to "defuse" these typhoons (reduce their destructive capacity), either by steering the center of the storm away from land or by preventing it from building up to its full strength and fury, all would gain—providing that the same quantity of water were deposited in the same places down the defused storm's track. Whether these goals—mitigating the damage while maintaining the rainfall necessary for agriculture—can be met simultaneously is not known but is worth ascertaining.

MECHANISMS FOR ATMOSPHERIC MANAGEMENT

What international measures could be taken, and what international organizations could be charged with what tasks? How do we get from here—with our currently rising but mostly isolated concern for the problems—to there—with an international will effectively mobilized for actions (or inactions) to benefit all?

One step, perhaps achievable fairly quickly, would be a ban on using the atmosphere for hostile purposes, including banning weather or environmental modification for those ends. The USSR and the United States, as noted earlier, already have developed draft materials along such lines for consideration. Currently, no country has the capability to create such effects while limiting them to the target country, yet all can appreciate the dangers should some country try to do so. Thus, such a ban is one from which all nations can benefit.

Another area, not under wide discussion but potentially similar to the above, would be an international agreement to ban modification of the Greenland and Antarctic ice sheets. These ice masses, along with the Arctic Sea ice, are coupled closely with the atmosphere and are integral parts of the thermodynamic system that drives the world's weather and climate.

One reason to formalize international understandings fairly quickly on these ice masses is that plans are extant for modifying them. One such plan is to sprinkle vast areas of Arctic and Greenland ice with coal dust. This would increase the amount of heat held in the system (because the white snow or ice reflects heat back out to space at a much higher rate than black dust would) and thus presumably counteract a possible global cooling trend; its advocates hope it will result in increased melting of the ice, opening northern ports to year-round shipping, and raising ocean levels (although this last would happen slowly—no tsunami would engulf the world's port cities in a few weeks, months, or even years).

Our current knowledge is not good enough to predict the exact consequences of a major modification of these ice masses, but is sufficient to suggest that the effect could be significant, both on the global average climate and in increasing local variability of weather patterns.

Perhaps, with increasing knowledge of how the ice masses interact with the atmosphere, we could learn to "fine tune" our weather patterns, offsetting a general cooling trend (if there is such) by one technique, a general warming trend by another. But those skills are well beyond us today. Rather like a small child trying to fix his Grandfather's fine old watch, the potential for unintentional harm far exceeds the potential for a happy improvement. The goal of such a ban on the modification of these ice packs would be to minimize disruption of the system until we learn how—and whether—to initiate such controlled changes.

Antarctica is currently covered by an international treaty which, among other provisions, prohibits sovereignty claims until 1990. Perhaps this treaty could be extended. Greenland is a different case; not only is there no international treaty on which to build, but it is claimed, and has been for centuries, as sovereign territory by Denmark. Would a treaty be either useful or achievable, or should we rely on the Danes' and Greenlanders' demonstrated concern for the environment to keep this area free from massive tampering?

A similar intersection of traditional sovereignty claims with newly perceived global effects keeps the Arctic ice pack from being added to this list of desirable areas to protect. As noted above, the Soviet Union is proceeding with its plans to divert several of its northward flowing rivers to the south. Within historical concepts of sovereignty, this is entirely their decision to make. With rising awareness of its possible impact on global circulation patterns, the case for sovereignty is much less clear. However, a realistic look at current world politics does not suggest that this is a fruitful time to raise this issue.

The World Meteorological Organization is generally recognized as an extremely effective information-gathering and disseminating institution. WMO projects are staffed almost exclusively with administrators and scientists, and there has been relatively little involvement of national policymakers in WMO programs or vice versa. Some meteorologists warn that to involve the WMO in international political questions would be to destroy its effectiveness by politicizing what would otherwise remain scientific questions. Others acknowledge this danger but feel that the base of expertise and cooperation is too valuable to be overlooked in reaching toward the desired end of a global forum on the atmosphere.

Acknowledging that any such global forum will only evolve slowly, what might be its mandate and organization? Four functional areas can be identified:

Policy: Determining which areas and questions the forum will address; delimiting the boundaries of national versus international responsibility.

Planning: Reducing programs to operational terms; defining research needed to carry out specific programs.

Administration: Carrying out of policies and programs adopted by the policymaking group.

Technical information: Gathering and disseminating data; carrying out research.

The first of these areas will involve governments in the range of political questions and disputes which are inevitably associated with the allocation of scarce resources. The common interests, however, in minimizing harm or threats of harm to a nation's economic and environmental interests are powerful centripetal forces. Such a forum would bring problems of the atmosphere to the attention of the world community, thus minimizing the risks of inadvertent or irreparable damage and raising questions of liability for damages before the fact.

There is today no analogous organization, encompassing both political and scientific questions. The WMO functions well as an administrative and information-gathering body; however, it has had limited experience as an operational organization responsible for carrying out specific experiments or programs. It is just beginning to move in this direction. An alternative to widening the mandate of the WMO to encompass all functions of a global forum would be to formalize and limit the WMO's roles by excluding the political and policymaking processes, and to establish a separate body to handle these. Such a body might be, for example, at the outset, an International Scientific Advisory Panel (which would encompass a number of ad hoc panels and could be under the auspices of the U.N. Environment Program or the WMO, with a joint working arrangement with the International Council of Scientific Unions) to assess the probable consequences of proposed large-scale modifications or, upon request, of other operations such as typhoon modification.

How long might we have to work out some binding international conventions for protecting the atmosphere, and thereby each other and ourselves? To date, the net impact of weather modification operations, from a global perspective, has been limited. Most of the operations have had either no success or they have contributed to increasing the favorability of local weather conditions. As long as this is true, states' interests in pursuing weather modification are consistent with their national interests in the weather system. During the next 10 years or more, however, we can expect that as weather modification projects become larger and more frequent, they potentially could have adverse effects on neighboring states or contribute to natural disasters, such as floods. We can expect increasing awareness and anxiety over military

applications of the technology, over potential damage from cloud-seeding operations or attempts to modify typhoons, or over the serious but largely uncertain effects from possible large-scale experiments. We may characterize these developments as the emergence of a threat of a "decreasing sum game" in the value of weather and climate to states.

The most important technical change from 10 to 20 years in the future probably will be a proliferation of weather modification operations which easily could have adverse effects on others, such as rain stimulation and typhoon modification. Weather modification operations probably will become more frequent and cover larger areas. It may be more feasible for states to initiate large-scale changes in the weather and climate. At the same time, we can expect greater scientific understanding of the consequences of weather modification operations, which will give states a clearer view of the benefits and costs to them of using the technology.

More and more situations are likely to arise in which one state's benefit is another's loss. The actual operations, however, will not be strictly constant sums, for the gains and losses transferred are unlikely to be equal, and the game will not have a symmetric matrix. At the same time, there will be some cases in which weather modification programs only contribute to more favorable weather, and potentially some in which they cause only worse weather conditions. Only at the aggregate level can we suggest that we may move to a period we can characterize as a dominantly "constant sum game."

However, during this period we also can expect a greater threat of a decreasing sum game. Adverse effects from weather and climate modification operations will become more common and occur on increasingly larger scales. Threats to the common interest of states in weather and climate will begin to emerge from the inadvertent effects of intentional modification programs, as from repeated hurricane modification or widespread cloud-seeding operations. Unless there is an agreement banning the hostile uses of weather and climate modification, we can expect many more countries to consider the techniques as potential weapons. The potential for adverse, but unforeseen, interactions and feedbacks in weather systems from intentional modifications will increase. It is likely to be impossible to control these effects once they are initiated.

The problems foreseen for the next few years (if not dealt with) will continue, and, in most cases, intensify during the following 20 years. In particular, weather and climate will become increasingly politicized, and disputes regarding meteorological conditions are likely to multiply. There will be increasing awareness of the impact of weather and climate modification on economic development, trade patterns, the ecology, and social behavior. Developing countries still will need technical assistance to establish their own scientifically based cloud-seeding operations or other modification programs.

Beyond 20 years we can anticipate much wider application of weather modification techniques and potentially the employment of large-scale techniques of weather and climate modification. Since weather modification programs will be common, we anticipate that the problem of adverse effects upon neighbors will arise frequently. The inadvertent cumulative effects from intentional weather modification programs over widespread areas and the cumulative effects in time from repeated modification programs should become apparent. Some weather modification operations are likely to affect weather in distant areas. Large-scale applications for military purposes of weather and climate modification techniques should become feasible, thereby threatening the value of weather and climate to many states. All of these will magnify the threat of a decreasing sum game in the value of the global climate and weather systems.

At the same time, we should have a considerably deeper understanding of weather and climate. As computer models for weather systems become more accurate and sophisticated and as computer capacity increases, it may be possible to predict more precisely the effects of a proposed modification. National attitudes toward programs will then be based on better information. By then we may have sufficient understanding of the feedback mechanisms in the global weather system to know whether we can increase the total amount of rainfall or whether efforts to increase rainfall will ultimately result only in the same equilibrium. Eventually, we may approach a situation in which—by the use of sophisticated computer models—we may be able to calculate the most favorable climate and even the most favorable weather that would be possible at any given time and place and the losses that these conditions would inflict on others. If our understanding of weather and climate systems should indicate that, by the appropriate application of modification technology we can continuously improve global weather and climate, we will be in a period with a potential for an increasing sum game.

The conditions projected beyond 20 years will make it necessary to have an international weather and climate authority to coordinate weather and climate modification programs, to ensure that states that could be affected adversely by proposed programs have consented, and to enjoin experiments or programs which could have severe adverse effects on others unless they had expressly consented. For large-scale operations, parties would be required to have an international scientific advisory panel, aided by a competent secretariat, assess the consequences, agree to abide by its recommendations, and, if a project proceeded, provide compensation to the extent feasible to those adversely affected. Most importantly, there will need to be processes developed for allocating opportunities to parties to improve their weather or climate. We will need procedures by which states can avoid a free-for-all resulting in worse weather for many countries or consistently making some states victims of the modification programs of others. This will require significant delegation of authority to an international body, beyond any that has been delegated so far or that is likely in the near term.

Whether the "near perfect" knowledge of weather and climate described above will increase or decrease acceptance by states of their community interest in the global climate and weather systems depends in a large part upon whether states see that the advantages of cooperating in using weather and climate modification techniques and the dangers of not cooperating outweigh the disadvantages of restricting their own operations. Past experience suggests that strong incentives for cooperation should emerge in this period. How-

> ever, if computer calculations show that certain states, because of their relationship to atmospheric circulation patterns or because of other factors, are more likely to be able to modify the weather in their favor than are other states, which may be consistently more vulnerable to weather modification operations of others, serious political conflicts could ensue. Efforts to observe a community interest in the global climate system could disappear.[2]

These are only suggestions of some ways to approach the problem of gaining international cooperation on what are truly international concerns. They might not be the best solutions available today; they might be even more seriously deficient in a decade or two, by which time many things will have changed. But they represent a place from which to start—a series of ideas for concerned individuals, organizations, and nations to consider and improve upon. Effective international cooperation has never sprung, Phoenix-like, to perfection in a day. But if we wait until the potential problems become current catastrophes, calls for effective cooperation will have the appropriateness of jokes at a dying man's bedside.

Just as a single cell is surrounded by a membrane, shielding it, permitting homeostasis to be maintained, and selectively admitting and rejecting nutrients and wastes, so the earth is enclosed by its atmosphere. Like the membrane of the cell, the "membrane" of the atmosphere is actively involved in the life processes it surrounds and is necessary to them. The organelles of the cell invest energy in the membrane in order to preserve themselves. Similarly, now that we recognize that we can damage this envelope that surrounds us, thus bringing damage to ourselves, it is time for humankind to invest some energy, forethought, and concern in our membrane, the atmosphere.

[2] This material was adapted from Edith Brown Weiss, International responses to weather modification. International Organization 29(3), and is based in part on the forthcoming book on the subject to be published by the University of California Press.

Part 6

The Atmosphere and Society

WILLIAM W. KELLOGG

INTRODUCTION

As stated in the Preface, the main purpose of this Conference is to anticipate the call that will be made on scientists and leaders of governments regarding the need to protect the atmospheric environment *before* these calls are made. A serious situation has already arisen in the oceanic realm, and we may avoid the kind of international altercations that have erupted over the Law of the Sea if we can foresee where similar problems may arise in the atmospheric realm.

Because the protection of the atmosphere and the need to protect ourselves from atmospheric vagaries are a common concern, it involves a new kind of mutual responsibility among all peoples of the earth, and it is therefore of tremendous importance to have sound scientific estimates as a basis for public action. That basis is what we have tried to provide in this report, and where we have not been able to finish the job, we have at least defined where most of the limits are.

This "Position Paper" (written before the conference) outlines briefly some of the things we know about the atmosphere and our concerns for the future of this domain. It also asks many questions designed to sharpen the discussion.

NATURE OF THE ATMOSPHERE, WEATHER, AND CLIMATE

Since our atmosphere is a restless fluid on a rotating earth, constantly driven by the heat from the sun and the loss of heat to space, its *motion* and its *changes* are its most apparent characteristics. These motions determine the patterns of weather; and the changes that occur seasonally, and also from year to year or from decade to decade, provide the statistics that we call "climate." The climate is determined by a complex and delicate balance in the total system of which the atmosphere itself is but a part—the atmosphere–ocean–land–cryosphere system.

It is well known that the climate has changed many times in the past. On a time scale of tens of millions of years, the present climate is considerably colder than the average, since for the past 500 million years there has been permanent ice around the poles only 10 or 15 percent of the time, as there is today. However, on a time scale of tens of thousands of years, we are now in a relatively warm "interglacial" period. In fact, the present interglacial period was established some 10,000 to 15,000 years ago, and if we believe that the quasi-periodic fluctuations of the last million years are going to continue, we should be due for another cooling trend that would last for 100,000 years or so—until the next interglacial.

The difficulty with any estimate of the future climate is that we are only vaguely aware of the *natural* causes of climate change, and we have no adequate model of the climate system that we can use to make a good forecast for a year, a decade, or a century. Such a model of the climate system would have to take account of the heat balance and circulations in the oceans, the behavior of the Arctic Ocean ice and the great ice sheets of Greenland and the Antarctic, volcanic activity that may produce stratospheric particles that screen the sunlight, changes in the solar radiation itself, and so forth. We know that there are many nonlinear interactions (where the behavior of the whole is not necessarily the sum of the contributions of the parts) within this system, often referred to as "feedback mechanisms," and our current climate models do not yet include all the important ones. Perhaps some day we will be clever enough to do so, and our computers will be fast enough to make all the necessary calculations, but we still have a long way to go. Is it too soon to ask what we would do if we did have a capability to forecast the future climate?

In spite of these shortcomings, however, we do understand enough about the atmosphere and its interactions with the ocean, the land, and the cryosphere to be able to say something about "what would happen if . . .?" That is most significant, because one of the burning questions being faced by mankind is: What will happen if we continue to change the face of the earth, the composition of the atmosphere, the heat balance of the continents, and so forth? These ingredients constitute the "boundary conditions" or external factors that help to determine the balance in the climate system, and our models can be of great use in estimating how a change in a given boundary condition will affect the climate of the globe or of a region, all other boundary conditions remaining constant—indeed, they are really the only tools we have to determine such things.

These matters will be treated further in the next section, and were discussed during the Conference. The important point to bear in mind is that *mankind surely has already affected the climate of vast regions, and quite possibly of the entire earth,* and that its ever escalating population and demand for energy and food will produce larger changes in the years ahead.

So far we have spoken mostly of changes that affect the whole earth. What about climate changes that affect a particular region, particularly one where agriculture is marginal and where a relatively small fluctuation has a devastating effect? We have witnessed the recent famine in the Sahel of North Africa, the failure of a wheat crop in 1972 in the Soviet Union, the delay in the start of the monsoon rains of India, the temporary failure of the anchovy catch off Peru, and recent floods in the midwestern United States that devastated an appreciable fraction of certain crops. **In each case, while the immediate effect may have been localized, the economic and social repercussions were worldwide. The nations of the world are becoming so economically interdependent that the failure of a major harvest anywhere sends shocks throughout the network of world trade and influences political decisions. No country can ignore the reality of climate fluctuations and climate change, nor their effects on the affairs of mankind. The question is: How do we cope with them in an interdependent world? This is a theme we will return to many times.**

Narrowing the time and space scale still further, we have daily evidence in the newspapers of the destructiveness of the weather, in the form of thunderstorms and tornadoes, hurricanes and typhoons, blizzards, hail, floods, high winds, lightning, and so forth. There is a curious human tendency to think of "normal" weather as fair weather and to view storms as "unusual," whereas nothing could be further from the truth. Where weather records have been kept for several decades climatologists can describe statistically the probability of a given kind of weather—the "hundred-year flood," for example, is the magnitude of the flood in a certain watershed than can be expected *on the average* once in 100 years. The same kinds of probabilities can be assigned (where we have adequate data) for almost any kind of weather, though they are seldom made use of by the public or even by our governments—dam construction generally being one exception.

The ability to predict where such destructive storms will occur for periods of up to 4 or 5 days has increased markedly as a result of a better knowledge of the processes that govern them, improved international observing networks and weather satellites, and high speed computers; and this ability certainly has saved many lives and much property. The most developed countries are those that can best afford to take advantage of modern advances in meteorology, and we have witnessed recently a tendency for other countries to combine their weather-forecasting efforts—for example, in Europe, East Africa, and parts of the Orient. In addition, the three World Meteorological Centers of the World Meteorological Organization (Washington, Moscow, and Melbourne) and the Regional Meteorological Centers share their hemispheric weather analyses and some forecast products. **Thus, in spite of many political barriers, there appears to be a gradual move toward a more international form of weather service. How far will this trend go? Will it be accelerated by the need for countries to take advantage of longer-term predictions, when (and if) these become possible? Or will such an ability be held by the countries that possess it as a strategic asset?**

MANKIND'S INFLUENCE ON THE ATMOSPHERE

In the previous section we mentioned the fact that mankind already has affected some of the boundary conditions of the global climate system, and will almost surely change them even more in the future. Mankind can also modify the weather deliberately in certain special situations, and if such endeavors were to become more intensive, their cumulative effect would be a climate change for the region involved. In this section we will trace some of these ideas and their implications a bit further.

Virtually all the artificial or anthropogenic changes that influence the climate do so through their effect on the heat (energy) balance of the climate system or some part of it. These are, in outline, the main ways in which this occurs:

1. *Changes of the land surface:* As cities expand, desert areas are irrigated, forests are cut down to make fields, grassland is overgrazed by cattle or goats, and so forth, the heat and water balance of the surface is altered. If the reflectivity of the surface (its albedo) is increased, less solar radiation will be absorbed and a net cooling of the surface of the region will occur. This, in turn, will have an effect on the stability of the atmosphere and its ability to form cumulus clouds, and there may thereby be a decrease in rainfall—a mechanism that has been suggested as the cause of the spread of such deserts as the Sahara and the Rajasthan. Similarly, the growth of cities can work in the opposite direction and cause an increase in rainfall immediately downwind. (Such observed increases are probably also related to the particles added to the air by the city.) Such changes of the climate are primarily local or regional, but their cumulative effect could have an influence on an even larger area.
2. *Adding to particle content:* Both industry and agriculture add particles (aerosols) to the lower atmosphere, as do automobiles, power plants, house furnaces, and so forth. An aerosol particle in the lower atmosphere has a mean lifetime of only 3 to 5 days and about one week at cloud level before it is rained out or washed out, but nevertheless the plumes of such manmade particles can spread for thousands of kilometers downwind from the industrial regions of the world. Their suspected effect on the weather and climate is twofold: They act as condensation nuclei (or perhaps freezing nuclei) and change the rate of growth of cloud droplets, usually enhancing rainfall or snowfall but in some circumstances decreasing it; and they also absorb and scatter solar radiation and thereby change the net heat balance of the atmosphere. Industrial aerosols probably cause a warming when they are over land (where about two-thirds of them are). Most anthropogenic aerosols currently exist in the middle lati-

tudes of the Northern Hemisphere, near or downwind from their main sources, and their distribution is very uneven. While there have been noticeable increases in the total particulate content of the atmosphere in industrialized regions since the early part of the century, there is evidence that these trends are leveling out in countries where air pollution controls are being enforced.

3. *Adding carbon dioxide:* Since the beginning of the Industrial Revolution, mankind has been taking fossil fuels out of the earth and releasing their carbon into the atmosphere as carbon dioxide. Roughly half of this added carbon dioxide has remained in the atmosphere, the other half being absorbed by the upper part of the oceans and taken up by the biosphere (mostly the forests of the world). The carbon dioxide content of the atmosphere was probably around 290 parts per million by volume (ppmv) when we started this process 150 years ago (though no precise measurements were taken then) and it is now over 320 ppmv. If we continue to burn fossil fuels at an increasing rate for the next few decades, as seems likely, it will rise to around 400 ppmv by A.D. 2000 and may double by about A.D. 2040. Carbon dioxide is a relatively stable gas and is quite well mixed throughout the atmosphere on a time scale of years. The effect of raising the carbon dioxide content is to warm the lower atmosphere and cool the stratosphere, since it is a gas that is transparent to solar radiation but absorbs some of the infrared radiation emitted from the surface that would otherwise escape to space—a process commonly (but probably incorrectly) labeled "the greenhouse effect." It is estimated, using our available climate models, that the global average warming at the surface from this effect alone will be nearly 1°C by A.D. 2000 and about 2°C by A.D. 2040. These estimates could be in error by as much as a factor of 2 or more in either direction, but even so a 1°C change would be large compared to most of the natural fluctuations that we have seen in the last 1,000 years. Changes in the mean surface temperature in the polar regions are expected to be 3 to 5 times larger than the global average change, based on both climate modeling and evidence from the real atmosphere.
4. *Adding heat:* The atmosphere is driven by solar heat, most of it the heat absorbed by the earth's surface. The average solar radiation absorbed at the surface is about 150 watts m^{-2}. (The total absorbed at the surface is about 8×10^7 Gwatts.) If all of the heat released by mankind were averaged over the earth, it would amount to a bit over 10^{-4} of this amount, or 0.015 watts m^{-2}, and this would have a trivial influence on the planetary heat balance. Suppose, as some have presumed, our population growth and per capita energy consumption were to rise until we were releasing 100 times more heat than we

are now, or an average of 1 percent of the solar energy absorbed at the surface—what then? Again we can resort to our climate models, such as they are, and find that this one percent increase in the average heat available to the climate system would raise the mean surface temperature about 2°C. Again we should place an uncertainty of at least a factor of 2 on this estimate, an uncertainty that is compounded by the fact that (unlike the carbon dioxide greenhouse effect) the patterns of this added heat will be spread very unevenly over the earth, probably being most where the people are. Incidentally, this amount of heat corresponds to a 20 billion world population (5 times the present), each person using 4 times the present per capita consumption in the United States, or 40 kw of thermal power (the world average per capita thermal power consumption is now about 2 kw). Is such a society possible? If so, when may we achieve it? It would be ridiculous to expect the current rate of growth to continue indefinitely, but where will society level off?

The conclusion that we must come to is that mankind is almost surely heating up the surface of our planet by adding aerosols, carbon dioxide, and direct heat. We may argue over the details of this picture, but the main direction we are taking seems rather clear. Some of the details are actually vitally significant, such as the influence that a global warming will have on the Arctic Ocean ice pack and the ice sheets of Greenland and the Antarctic. Will the Arctic Ocean become ice-free as the polar regions warm up? Will the ice sheets melt or grow? (There may be more snow on their tops to counteract the more rapid melting and ablation at their edges.) What will happen to the mean sea level and the coastal cities around the world? Unfortunately, there are many more possible scenarios of the future than there are firm predictions.

If we are due for an unprecedented (in the history of civilized man) global warming, what actions should the nations of the world take, either separately or in concert? Is there any international mechanism in existence now that can adjudicate between the claims of the losers and the rest of the world, or make long-range plans to guide the many societal readjustments that will be required? What would such an organization or mechanism be like, what would its powers be? (The need for such international mechanisms was stressed in the Preface and are discussed at length in Part 5.)

So far we have been dealing with *inadvertent* weather and climate modification, but the time is approaching when *purposeful* weather modification may take place on a larger scale and begin to have effects that extend beyond national boundaries. And what about more grandiose schemes to modify the climate of a region?

Though progress in *weather* modification has generally been slow, and the state of the art often has been oversold by those who make a living by it, there are some areas where progress has been made. For example, cold fog dispersal, augmentation of snow pack in mountains, suppression of hail, and increasing rainfall under very special conditions all seem to have been demonstrated with modest success, and there is hope that we will some day be able to modify or steer hurricanes. This growing ability to modify certain types of weather raises a new set of problems where the people involved cannot agree on the way they want their weather to be modified. While the litigation so far usually has been a domestic matter, it may move into the international arena some day. Do we have any international mechanisms to cope with such disputes when they arise? The World Meteorological Organization (WMO) has just this year agreed to create an international Panel of Experts to advise less developed countries on weather modification and to help in carrying such programs out, and perhaps this is a modest step in the right direction.

Schemes to modify the *climate* on a large scale have been suggested many times, sometimes seriously, though so far they generally have been considered as on the fringe of science fiction. If one imagined a substantial fraction of the resources now devoted to armaments being devoted to such grandiose projects, then some of them would seem feasible—such as removing the Arctic Ocean ice pack by damming the Bering Strait and pumping water or diverting the rivers that flow northward into the Arctic Ocean (being considered seriously in the Soviet Union), or creating a dust cloud in orbit that would shield the earth from the sun, or diverting hurricanes by changing the ocean surface temperature over many square kilometers. The list of suggestions is already fairly long, and no doubt more will be added. (Fig. 2 is a summary of these ideas.)

The fatal objection to any such scheme is, of course, that we do not know enough about the climate system to be able to predict all the consequences of tampering with it. Thus, we might be causing a large disruption, possibly "irreversible," and it is difficult to predict who would benefit and who would lose. It might even turn out that everyone would be losers.

Suppose, however, that progress in understanding and modeling the climate system does progress to the point where reasonably reliable predictions of the effect of a climate modification project could be made. Would we then wish to go ahead with it? Who would make the decision? Who would see that the losers were compensated properly? Again, there is no international mechanism now that could deal with such a set of problems, though perhaps it is not too early to start to organize it.

To summarize: **The atmosphere, an important part of the climate system, is subject to natural short-term fluctuations and longer-term changes that influence all our lives. In an increasingly interdependent**

society, a crop failure or natural disaster in one place has repercussions that cross national boundaries. Now mankind, by both purposeful acts and inadvertently altering the atmospheric boundary conditions, has induced additional climate changes, and these probably will become very apparent in the next few decades. The main effect, so far as we can predict now, will be a major global warming, a warming that will be felt most strongly in the polar regions.

Do we have any national or international mechanisms that can help society in its readjustments to both natural and manmade climate changes? Should we begin to design such mechanisms now? (These are among the many questions discussed at the Conference.) Above all, scientists will have to decide on what the minimum hazards and readjustments for us will be, since they will be the basis for society's response.

ATMOSPHERIC POLLUTION AND AIR QUALITY

Air pollution is a currently recognized problem in nearly every large city. That air pollution can drift across national boundaries is also evident, and now we are forced to realize that even the global atmosphere can be modified by the products of mankind's activities. Thus, whereas the preservation of air quality in the past has been considered as a local problem, that is no longer the case.

The Conference considered the matter of atmospheric pollution and air quality from a number of viewpoints, with emphasis on the international measures that might be required in the future if the situation continues to become more serious. How "serious" are the changes that have already taken place? What are some of the problems that may arise in the future?

In the previous section we have touched on some of the *climatic* effects of the contamination of the atmosphere on a large scale, notably the increasing carbon dioxide content and the possible regional effect of particles in the air. The reason carbon dioxide can build up in the atmosphere is that it is a stable gas in the troposphere and stratosphere, and its removal by the ocean and biosphere is on a longer time scale than mankind's rate of addition, which is increasing exponentially with a doubling time of 20 to 30 years. Most other things we release in the atmosphere, such as particles, sulfur dioxide, carbon monoxide, oxides of nitrogen, volatile components of petroleum, and so forth, are not so stable chemically or are subject to rapid removal by rainout and washout. Thus, as we pointed out earlier, aerosol particles have a mean lifetime in the lower atmosphere of less than one week, and many of our pollutants (notably sulfur dioxide and most volatile petroleum products) are converted to particles and removed by the same mechanism.

The fact that sulfur dioxide gas and the sulfate particles that are formed from it do not remain in the atmosphere indefinitely does not preclude their influence being felt far downwind from their source. With a mean lifetime of 5 days, for example, and a typical mean wind near the surface of 5 m/sec, the average distance traveled by a particle before its removal is nearly 2500 km. This is why such products of an industrial society are detected far out in the western Atlantic and in relatively remote continental areas where there are no sources nearby.

Attention has been drawn to the situation in Scandinavia, notably in Sweden, where there are "acid rains" that have raised the acidity of Swedish lakes and rivers to a point where the fish there are endangered. The sulfate that accounts for this unwanted importation to Sweden comes from Britain, Germany, Poland, and the Soviet Union, depending on the direction of the wind. The Swedish Royal Ministry for Foreign Affairs and the Royal Ministry of Agriculture published an account of its findings, entitled, *Air Pollution Across National Boundaries: The Impact on the Environment of Sulfur in Air and Precipitation,* and this was presented at the U.N. Conference on the Human Environment, held in Stockholm in 1972. One of the conclusions of this conference was as follows:

> Principle 22: States shall cooperate to develop further the international law regarding liability and compensation for the victims of pollution and other environmental damage caused by activities within the jurisdiction or control of such States to areas beyond their jurisdiction. . . .

and then:

> Principle 24: International matters concerning the protection and improvement of the environment should be handled in a cooperative spirit by all countries, big or small, on an equal footing. . . .

Although the Swedish representatives have been unable so far to obtain any satisfactory resolution of their complaints from the United Nations or the World Court, on the other side of the world another international air pollution dispute was settled in 1941 by arbitration in a court of law. The Trail Smelter in British Columbia, Canada, had been emitting sulfur oxides and this was causing agricultural damage across the border in the United States. The case was taken to a Tribunal (jointly established by Canada and the United States), and the outcome was a cease-and-desist order to the Canadian Pacific Railway (owner of the smelter plant) and an award of damages to the aggrieved farmers across the border.

The case of the Trail Smelter is an example of arbitration between two neighboring countries with generally cordial relations with each other. The United States and Canada, incidentally, also have set up a bipartite commission to reduce water pollution in the Great Lakes and the St. Lawrence River, bodies of water whose banks are shared by the two countries.

So far, no such case has been tried by the World Court in the Hague, though there have been attempts to bring them to that Court, e.g., Australia's and New Zealand's unsuccessful attempt to force France to stop its nuclear testing in the South Pacific by legal action.

These are examples of rather limited intrusions of the pollution of one country into the air space of another, but there are now examples of a worldwide activity shared by many nations impacting the global atmosphere. In addition to carbon dioxide, which will be causing a global warming in the years ahead (discussed in the previous section) oxides of nitrogen released into the stratosphere by high-flying (supersonic) aircraft and nitrate fertilizers may cause a global depletion of stratospheric ozone. The continued widespread use of chlorofluoromethanes ("freons") in spray cans and as refrigerants also may deplete the ozone layer. Since the ozone layer screens out solar ultraviolet radiation, a decrease in the total ozone amount in the stratosphere will let more ultraviolet through. It is not yet clear what effects this would have on biological systems, including mankind, but a commonly accepted belief is that the increased ultraviolet radiation would result in more cases of skin cancer among the Nordic populations, and some ecosystems (including crops) may be adversely affected—though so far there is no firm evidence to support the latter supposition.

In this kind of situation action by one country can do little to change the course of events. Supersonic aircraft will be flown by many countries (if they prove to be economical to operate), nitrate fertilizers are in widespread agricultural use nearly everywhere (and they will be required in increasing amounts to sustain a growing world population), and the industry supplying chlorofluoromethanes (only about half of which is in the United States) is expecting to increase steadily its rate of production.

While the environmental impacts of supersonic transports seem to have been of great public concern only in the United States, the threat posed by the chlorofluoromethanes has attracted attention in many countries. As evidence of this, the World Meteorological Organization, at its Seventh Congress in Geneva in May 1975, adopted a number of resolutions relating to "WMO activities in the field of environmental pollution," and agreed that there was "an urgent need for more studies (and for a definitive review of these studies) to determine the extent to which manmade pollutants might be responsible for reducing the quantity of ozone in the stratosphere." The WMO Commission for Atmospheric Sciences convened a Working Group on Stratospheric and Mesospheric Problems to consider these matters in September 1975, and some of the results of that meeting were reported to this Conference (WMO 1975).

While the WMO has responded to the concern over stratospheric pol-

lution with an effort to determine first the facts of the matter—a sensible step in any case—it is recognized that this international body has no authority to take any action, if action seems to be called for. Large industrial concerns, the aircraft industry, and agricultural interests will probably all try to oppose any international efforts to limit their operations, especially if they can show that such limitations are premature. We have already heard samples of their counterarguments, which are to the effect that the environmental effects of their products really are not understood adequately at this time, and that these effects are, in any case, so trivial compared to other interventions of society that they deserve to be ignored. The obvious inconsistency of this argument apparently does not weaken its soothing effect on the public—a phenomenon discussed in the Preface, in which people tend to forget the pain of past disasters and tend to deny warnings of hazards to come.

Perhaps no better example could be found to illustrate the need expressed by Margaret Mead in the Preface, to "now deal with a domain that must either be shared and responsibly protected by all people or all people will suffer."

APPENDIX I

The Carbon Cycle and the Paleo-Climatic Record

WALLACE S. BROECKER
Summarized by J. Dana Thompson

INTRODUCTION

The following remarks will address the CO_2 problem, which has long-range implications for our atmosphere and climate. The strategy for presentation is to review first some basic facts about the carbon cycle and the paleoclimatic record, about which scientists are in fairly good agreement, and then to address the more controversial question of what those records may imply about past and future climates.

A Review of the Atmospheric CO_2 Problem

If all known coal reserves of the earth were combusted, the CO_2 content of the atmosphere would increase by a factor of from 5 to 10. Man clearly has the *capability* of altering significantly the natural CO_2 content of the atmosphere. Of the possible climatic effects induced by man, only that for CO_2 can be demonstrated conclusively to be globally significant. Ekdahl and Keeling (1973) have shown from measurements on the island of Hawaii that the CO_2 content of the atmosphere rose an average of 0.7 parts per million (ppm) per year from 1958 to 1972. Theoretical back calculations indicate that the partial pressure of CO_2 in the atmosphere has risen about 15 percent since the dawn of the Industrial Revolution, from about 285 ppm to about 328 ppm. Projections of increased fossil fuel usage imply that the atmospheric CO_2 content will increase at a global average yearly rate of roughly 3 percent. The exact rate of increase is extremely difficult to determine because of the uncertainty in the predictions of fossil fuel usage, which in turn depends on population growth, increased use of alternative energy sources, and other factors. It appears certain that atmospheric CO_2 content will continue to increase over the next half century. It appears likely, based on present estimates, that a doubling of the current atmospheric CO_2 content will occur sometime in the next century (Bacastow and Keeling 1973; Broecker 1975). (See Fig. 4.)

If all CO_2 generated from the burning of chemical fuels remained in the atmosphere, the rate of increase in atmospheric CO_2 content should have been about 1.5 ppm per year in recent years. Apparently, about half the CO_2 added to the atmosphere is removed to the sea and to the terrestrial biosphere. Broecker et al. (1971) have calculated that uptake of CO_2 by the sea may account for 35 ± 10 percent of the CO_2 produced. Other investigators have arrived at lower percentages of oceanic uptake. If indeed the oceans (rather than the terrestrial biosphere) do take up most of the added CO_2 not retained in the atmosphere, it appears that the fraction of CO_2 produced by mankind and remaining in the atmosphere will not change significantly over the next several decades.

The oceans potentially are capable of taking out roughly 80 percent of the excess CO_2 produced by the burning of *all* fossil fuels. As the ocean acidity increases by CO_2 uptake, the ocean sediments would be attacked. This would reduce the acidity and permit more uptake of CO_2. However, the time scale for the "eating away" of bottom sediments is extremely long, on the order of thousands of years. There is little doubt that, given enough time, much of the excess CO_2 in the atmosphere could be taken up by the oceans in the form of bicarbonate ions. However, the time required is so long (at least 1,000 years) that the atmosphere probably is "programmed" into a higher CO_2 content over the next century.

Climatic Consequences of Increased Atmospheric CO_2 Content

Having established the virtual inevitability of increases in atmospheric CO_2 content over the climatically short time scale of a century, one is drawn to the question of how such increases will affect the climate and man's activities. A number of groups has estimated the change in global temperature that would result if the atmospheric CO_2 content were to double. Our best estimates come from globally averaged models of the atmosphere and experiments with three-dimensional numerical models of the atmosphere, run on large digital computers. Although these models include many physical processes occurring in the real atmosphere, even the three-dimensional models remain rather primitive when compared to the enormously complex system they are designed to simulate. Nevertheless, independent estimates from a handful of such models suggest that a doubling of the CO_2 content of the atmosphere would result in a rise in the global temperature of from 0.8°C to 3.6°C, with a value of about 2.5°C being widely quoted (Manabe and Wetherald 1967, 1975; Schneider, 1975). **Changes in surface temperature due to changes in atmospheric CO_2 content are not uniform over the earth. Both theory and observation suggest that they may be amplified in the polar regions. One might expect to observe first the climatic effects of man's changing the atmospheric CO_2 content at high latitudes.**

In view of the assertion that a doubling of atmospheric CO_2 content could raise the global average temperature of the earth's surface by 2.5°C, a whole series of questions arise.

Why has the global temperature been falling over the past several decades if increased atmospheric CO_2 content is predicted to warm the earth? What effect would a 2.5°C temperature rise have on the climate and the biosphere if it should occur? Is the climate stable to such perturbations in global temperature or is there a climatic "precipice?" If the climate is stable, does it "oscillate" like a plucked string when perturbed? The remainder of this contribution is addressed to those questions.

The mean global temperature has been falling over the past several decades, leading many observers to discount the warming effect of the CO_2 produced by the burning of fossil fuels. The temperature records shown in Figs. 9 and 10 strongly suggest that the present cooling is one of a long series of similar natural climatic fluctuations. It is possible that this cooling has temporarily more than compensated for the warming effect produced by burning chemical fuels. As the present natural cooling ceases during the next decade or two, the CO_2 warming effect could become significant. If this argument is correct, the first decade of the next century could mark the warmest global temperatures experienced in the last 1,000 years, as shown in the figure (Broecker 1975). While there is great uncertainty in obtaining such estimates, it is clear that we cannot yet discount the effect of CO_2 increases on the global climate.

Assume, therefore, that a 2.5°C warming of the global mean surface temperature is a reasonable estimate, given a doubling of atmospheric CO_2 content—what then? What effect would such a temperature rise have on the climate and biosphere? Perhaps it is best to avoid the direct approach of using models of environmental change to assess the climatic and biotic impact. Rather, we should examine the natural record of past climatic change. The part of the record best suited for our studies is a description of the environment that existed 18,000 years ago at the height of the last glacial period. John Imbrie and the NSF-sponsored CLIMAP Project have reconstructed an environmental "map" of the conditions that existed during full glaciation, mainly using information deduced from deep-sea cores. It appears that the average global temperature was about 4°C colder than the global temperature today. Regionally, it appears that there was very little temperature change in equatorial latitudes between glacial and interglacial periods, while there were large (greater than 6°C) changes at high latitudes.

We have quite good information on the biological and environmental conditions on the continents during the last glacial period. We know that they were profoundly different everywhere, except perhaps in the wet tropics. In many ways, as far as human activity is concerned, the planet is less hostile now than 18,000 years ago. Much more land is suitable for agriculture now than it was during the glacial period. For example, central Europe and western Russia were probably steppe country and tundra 18,000 years ago.

A 2.5°C temperature change cannot be ignored in light of the drastic environmental differences attendant to a 4°C temperature change between glacial and interglacial periods. We must consider seriously the impact of such a change over the short span of a century. It may be that such a change is beneficial rather than harmful to man. Perhaps we can grow more food due to warmer weather. In any event, the environmental consequences of such a warming may profoundly affect man's activities and his quality of life.

The question of climate stability is, of course, extremely important. Is our present climate stable to a "sudden" perturbation of 2.5°C? What about large amplitude oscillations of the climate following such a major perturbation? The past climatic record provides some clues as to how the climate varied during and following the last Ice Age. Figure 24 depicts an index of the size history of closed basin lakes in various parts of the world, based on geological records. This index is a good indicator of aridity, representing the combined effects of evaporation, precipitation, and basin drainage.

Using radiocarbon dating methods, the absolute chronology of the Great Basin of the Western United States has been deduced by Broecker and Kaufman (1965) (shown in upper left of Fig. 24). Note that the entire pluvial period came during the *decline* of the Ice Age. Further,

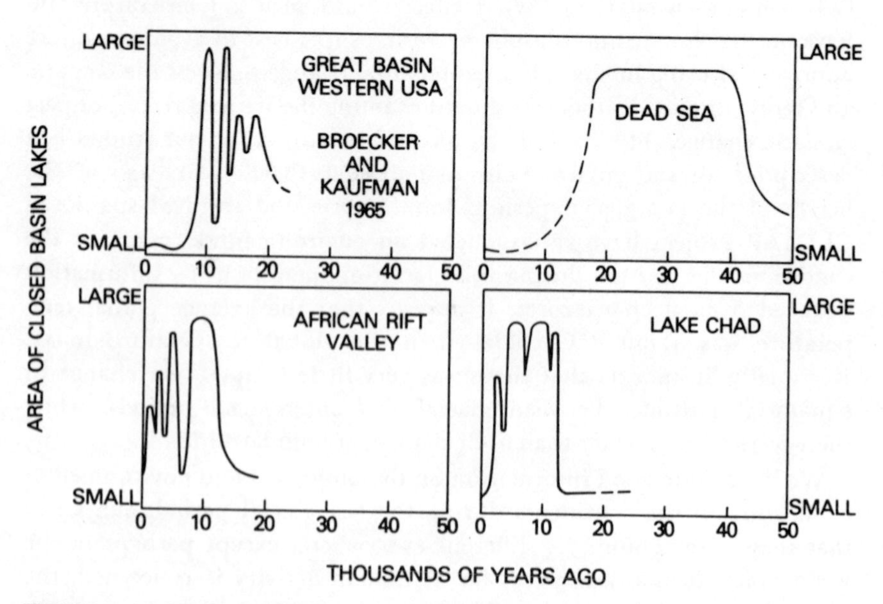

FIGURE 24. The size history of closed basin lakes in various parts of the world, based on geological records.

there were very pronounced variations in basin size following the glacial period. About 11,000 years ago, the lake went from its highest level to almost the present level and back to its highest level in a period of 200 to 500 years. (Minima to maxima in basin size represent a threefold increase in area of the lake.) This phenomenon of large, short-period variation in aridity has also been observed in the Greenland ice cores. Clearly, there is a strong suggestion that during the transition from glacial to interglacial periods there were large oscillations in climate. We have no idea what these oscillations were due to—but it is an indication that sharp changes in earth climate may be followed by large, short-period climatic oscillations.

Figure 24 also indicates that the response of desert lakes to climate change is not simple. For instance, Lake Chad's pluviation began at the end of the last glacial period and extended into the present interglacial period—with large fluctuations in basin size appearing immediately after the end of the Ice Age. The Dead Sea, however, had its pluvial period during the peak in glaciation. In other desert areas the pluviation maximum appeared after the glaciation maximum. It is not obvious what effect at 2.5°C global warming would have on precipitation patterns in light of these historical records.

In conjunction with the evidence for climate oscillations, one wonders if there is a climatic "precipice," a threshold past which the natural climate is perturbed so strongly that rather than returning to its old equilibrium state, it moves to a new, dramatically different state. Such climatic instability is advanced as one explanation for the sudden onset of glacial periods.

The evidence from the theoretical and numerical models is inconclusive about the stability of the earth's climate. Certainly our models are far too primitive to assess whether a 2.5°C global warming would push the climate toward an instability. One problem is that many of the "feedback loops" which are present in the real atmosphere are grossly oversimplified or totally omitted from the numerical models.

The geological evidence clearly indicates that the Ice Ages were very different climatically from interglacial periods like the present one. However, except for changes in solar constant, which we know very little about, no mechanism (e.g., volcanic dust, changes in atmospheric CO_2 content or orbital elements of the earth) has been proposed which is of sufficient strength to *directly* perturb the climate from interglacial to glacial conditions. We must conclude that if the solar constant did not change, the observed climatic effects were produced by positive (amplifying) feedback effects between the atmosphere, ocean, and cryosphere.

There is hope we will be able to use known, slight changes in the orbital elements of the earth to "calibrate" changes in the radiation distribution over the earth during the past 100,000 years. Thus it will be possible to assess the response of the climate to those changes based on paleoclimate records. The orbital changes modify the seasonal distribution of radiation but do not affect the annual radiation input to the earth as a whole. The periods of these changes, roughly 20,000 and 40,000 years, have now shown up in sea level heights deduced from coral reefs and in spectral analysis of temperatures deduced from ocean cores.

In summary, we recognize that mankind does have the potential for significantly altering the climate. We know that in the past major changes in climate have occurred. These changes were accompanied by dramatic changes in the global biological ensemble. We do not know why these changes occurred and further, we presently do not have the theoretical capability to predict what future climate changes might occur in response to man's activities. Decades of vigorous research lay between our present knowledge and more reliable assessments of the potential for manmade and natural climate change.

Comments and Discussion

Dr. Broecker's presentation evoked a score of mostly technical questions concerning the observations of atmospheric CO_2 content; his clarifications are incorporated in the above narrative. A recurring comment concerned the probability of doubling the atmospheric CO_2 content in the next century. It was pointed out that the doubling time depended crucially on the estimated rate of growth of population, energy consumption, and the fraction of total energy produced by chemical fuels.

During the discussion of the paleoclimatic record and changes in global temperature, skepticism was expressed concerning how representative temperature data from a single geographical area are for the globe as a whole (see Fig. 24). Dr. Broecker admitted the cycles observed in the Greenland ice cores did not appear in cores from other regions, but defended his interpretation by pointing out how large the amplitude of the signal was in the Greenland core, citing the theoretically predicted signal amplification effect at high latitudes.

During the discussion of a 2.5°C global temperature increase, it was emphasized by both the speaker and participants how primitive and limited our present numerical models of the atmosphere really are. The lack of an interacting ocean circulation, extensive omission of feedback effects, and the inability to resolve important scales of motion were several deficiencies mentioned.

A handful of participants expressed concern about local climatic changes introduced by direct thermal pollution as produced by large power parks and cities. While it was noted that global effects of direct heating are only a small fraction of the effect produced by atmospheric CO_2 increases, it was acknowledged that local direct heating provided a dramatic example of man's ability to alter the climate of his immediate environment.

The question of rising sea levels produced by global warming of the ice sheets of Greenland and the Antarctic was mentioned by the participants. It generally was agreed that sea-level rises would be more an expensive annoyance than a catastrophe. The large amount of energy required to melt ice implies that sea-level rises of only a few centimeters per year (at most) could be expected under a 2.5°C warming.

A point continually discussed by speaker and participants alike was that our knowledge about past climate, climate change, and man's effect on climate is extremely limited. A number of participants underscored the difficulty and uncertainty of making climate predictions and emphasized that in the face of our gross ignorance about climate change, certain policy and decisionmaking processes must proceed from incomplete and uncertain information. It was pointed out that scientists and policymakers must interact, and that the scientist must communicate clearly both his level of understanding *and* his level of ignorance of the decisionmakers.

APPENDIX II

The Interaction of the Atmosphere and the Biosphere

JAMES E. LOVELOCK
Summarized by J. Dana Thompson

A speculative and currently not too popular hypothesis is that the atmosphere might be a homeostatic system regulated by the biosphere. I can trace my own ideas back to childhood—my father was a typical Englishman, a rather sentimental gardener, and I can recall that he rescued wasps that were in danger of drowning in the water bucket in our garden. He always used to justify this kindly act by saying, "They're there for a purpose, you know." I guess his notion was that God had put the things in the world for our benefit and that they had a purpose. Now I suppose he would go along with the notion that they are part of our planetary life-support system and therefore are not to be destroyed because they might be useful to us.

These sorts of humanistic (mankind-oriented) world models are not very helpful if you want to do scientific tests on the system—whether they are right or wrong, they certainly do not help so far as the experimental scientist is concerned. There was an older, classical world model, an alternative to the Christian and humanistic one, which is very much more open to tests. In this classical model the whole world is seen as a living entity in which we, as a species, are seen more as a part, or partners, than as possessors. In deference to the earliest literature citation on this, we call it the "Gaia Hypothesis" after the classical Greek word for "Mother Earth." It is a more convenient term than "Biological Cybernetic System with Homeostatic Tendencies," which some have used. The term "Gaia" henceforth will refer to that system. (This concept originally was advanced by Lovelock and Margulis in 1974.)

To reintroduce an ancient hypothesis that the world is a dynamic biophysical system needs a bit of justification. I will try to highlight briefly the existing evidence to justify it, at least putting it forward as a sort of washline on which to hang experiments and ideas. First, the chemical composition of the atmosphere, the oceans, and the crust are very profoundly departed from the expectations of a steady-state chemical equilibrium. This is particularly true in the atmosphere, where the presence of gases such as methane, nitrous oxide, and even nitrogen in an oxidizing atmosphere violate rules of chemistry by tens of orders of magnitude. Disequilibria on this scale, particularly when one remembers one is dealing with a fluid medium maintained on an ongoing basis for

hundreds of millions of years, suggest that the atmosphere is more than just a geochemical product but might possibly be a biological contrivance somewhat like the paper of a wasp's or hornet's nest, which is there to maintain a chosen environment.

The second body of evidence is that there is a large set of planetary properties which are, and apparently always have been, just close to optimal for life. Thus, the surface temperature and the sea salinity and its pH have not changed from their present levels appreciably throughout the history of life on the planet, and in the past few hundreds of millions of years the oxygen concentration has remained similarly at the present safe level of 21 percent. This sort of constancy has been maintained in the face of probable large changes in solar output. (It is thought that the sun has increased its output between 20 and 50 percent during the time that life has been on the earth. This is quite a large change compared to the more recent changes discussed by Dr. Broecker.)

Perhaps everything has always been for the best by chance, or life adapted to an ideal planet, or maybe God has kept it so, but we propose self-regulation as an alternate hypothesis. In the following table are listed some of the gases of the atmosphere to illustrate the arguments just cited. Here we have nitrogen, oxygen, methane, nitrous oxide, and ammonia:

GAS	ABUND-ANCE Mol Fraction	FLUX Terramol Yr.$^{-1}$	EXPECTED ABUND-ANCE Mol Fraction	ENERGY (%)	HYPOTHETICAL FUNCTION
N_2	0.8	0.36	10^{-10}	3.6	Inerting gas, pressure builder, sink for nitrate
O_2	0.2	1670			Reference gas
CH_4	10^{-6}	60	10^{-35}	4.0	Ventilation of anaerobic sector, oxygen regulation
N_2O	10^{-7}	14	10^{-13}	1.2	Oxygen and ozone regulation
NH_3	10^{-8}	88	10^{-35}	1.3	pH regulation, greenhouse gas

From Lovelock and Margulis (1974).

The first column shows their abundance, which goes from nearly 100 percent in the case of nitrogen, down to somewhere in the region of a part per 10^8 in the case of ammonia. Not very long ago, gases such as methane, nitrous oxide, and ammonia would have been considered unimportant traces because of their small abundances, but this sort of view of trace gases in the atmosphere is belied by the second column which shows their flux through the air in units of 10^{12} moles per year. Leaving aside oxygen which fluxes at a very substantial rate, all the rest flux at very similar rates through the atmosphere, ammonia fluxing at least as fast, and probably *more* rapidly, than does nitrogen. In that sense they are comparable in significance.

The next column shows the expected abundance if thermodynamic and chemical equilibrium were the determining factor. All of them, even nitrogen, show absurdly large orders of magnitude departure from the expectations of equilibrium chemistry. (I do not think that 10^{-35} for methane and ammonia mean anything significant—it was just the limit of precision of the computer used to model the presence of methane at a level that is vanishingly small.) When one allows for the probability that methane has been around at 1.5 parts per million for tens or even hundreds of millions of years, the improbability that it would be there from inorganic sources or by accident reaches an extraordinary level.

The next column shows the fraction of solar energy required by the biosphere to cycle these gases through the atmosphere. For all of them it is well over one percent. One of the properties of biological systems is their frugality and parsimony about waste—to use such a large fraction of the total energy (the total sum nearing 10 percent) for cycling either a trace gas or an inert gas such as nitrogen suggests that in the context of our hypothesis they might have a function in the atmosphere.

The last column in the table considers the hypothetical function of each of these gases. Considering nitrogen first, an interesting possibility is that it might act as an inerting gas. As we will see, an atmosphere composed of 100 percent or nearly 100 percent oxygen is not permissible as far as life on the land surfaces is concerned. The next possible function is as a pressure builder. It may be that in the course of time meteorologists may demonstrate that the present 1,000 millibar atmosphere is the one that gives the desirable climate and if nitrogen were not present to maintain the pressure, it might be difficult to maintain the climate we have. The last, and perhaps the most definite, function for nitrogen in the atmosphere is as a "sink" for nitrate ions in the sea. The thermodynamically stable form of nitrogen is the nitrate ion dissolved in the sea. And if it were in that form, the sea would be about 0.2 molar more salt than it is, which would raise the total salinity of the sea to a point which is beyond that tolerable for most living creatures. Also, that much nitrate might be toxic in its own right, so there is a good biological reason for transferring nitrate from the sea to an inert form such as nitrogen gas in the atmosphere.

Oxygen, of course, is the redox reference gas against which all others are compared. We will say more about oxygen below.

The next interesting gas, methane, has two quite intriguing hypothetical functions. The first is as a ventilator of the anaerobic sector of the planet. It is produced in the anaerobic muds at the bottoms of marshes, ponds, lakes, rivers, and the continental shelf. As it bubbles up through those zones it carries with it, away from that region, poisonous materials such as oxygen itself, and the volatile components of elements which are toxic, such as lead, mercury, arsenic, and their metal derivatives. Methane might then have the function of maintaining the health of the anaerobic world as a ventilating gas. An equally if not more important function of methane is as an oxygen regulator. It is commonly agreed that the current oxygen level on the earth comes from a balance between the burial of carbon at a rate of about 10^{14} grams per year in sediments, giving a net increment of oxygen to the atmosphere, and the oxidation of reducing rocks exposed by weathering or of reducing materials released from volcanoes. Therefore, there is a balance between burial, which is undoubtedly a biological process, and the inorganic process of removal. The time constant of this process is a million years or so. It so happens that if the oxygen concentration is raised by as little as one percent, flammability (ability to support combustion) nearly doubles, and by the time one reaches the range of 30 to 35 percent flammability is nearly as great as in pure oxygen itself. It is very easy to demonstrate by experiment that somewhere in the region of 30 to 35 percent oxygen even wet vegetation will ignite by an electric spark or any other ignition source and burn completely. If we had 25 to 35 percent oxygen on the planet, standing vegetation of all sorts would be impossible and life would exist only in ponds and lakes. If methane were not released from the anaerobic sector, then presumably that much more carbon would be buried. It so happens that the carbon equivalent of the current methane production (about 2×10^{15} grams per year) is about 20 times the burial of carbon. Therefore, if methane were not produced, the oxygen concentration would rise about one percent per 2,500 years, which would lead to the disastrous conditions I mentioned in a few tens of thousands of years. This suggests an interesting and important function for methane.

Nitrous oxide might be connected similarly with oxygen regulation, but in the opposite sense, because it is a gas in which oxygen is sequestered and not really available for biological use and which can come to the atmosphere from the other sedimentary zones. It might also be concerned with ozone regulation in that when it reaches the stratosphere, where it decomposes, its products are potential ozone depleters. It is conceivable that too much ozone might be as unpleasant as too little!

Finally one comes to ammonia. One obvious purpose for ammonia could be pH regulation (pH is a measure of acidity or alkalinity of water). It is very easy to calculate that if ammonia were not made biologically, then the rainfall throughout the world, which is nearly neutral and has a pH of roughly 6, would be somewhere in the region of pH 3 to 4 (more acid) from the natural oxidation of nitrogen and sulfur compounds in the atmosphere. (I do not refer to anything man-made being put into the atmosphere.) Thus, the whole world would be facing the "acid rain" problem that Sweden and other European countries face at the moment. This seems to be an important function for this gas. Carl Sagen has suggested that in earlier times, before oxygen was in the atmosphere, ammonia might well have served as a useful "greenhouse" gas, keeping the planet warm during times when the sun's output was lower than it is now. It certainly is a more efficient infrared absorber than is CO_2—and the production of ammonia at current rates in a nonoxidizing atmosphere might well have led to a sufficient concentration to serve this function.

The Gaia is still a hypothesis. The facts and speculations just given are only a part of our collection, but the whole of it only corroborates—it does not prove—the existence of Gaia. But let us continue to assume that she does exist and see if this hypothesis has any useful bearing on our current problem of man and the atmosphere.

The first thought that comes to mind in this connection is pollution. Without in any way wishing to deny its importance to us as a species, to the planet as a whole it may be much less important than we think. We tend to forget that pollution is the way of life of many natural species and was so long before we appeared on the scene. Substances such as tetramethyl lead, dimethyl mercury, and trimethyl arsene have been dumped by anaerobic microflora into the oceans for hundreds of millions of years, their way of disposing of poisonous wastes of the anaerobic world. Perhaps the greatest air pollution the world has ever known was the emergence of oxygen itself; when this happened, whole sets of species must have been driven underground never to return to the surface, and others were destroyed. One has only to imagine a new marine system somehow able to produce chlorine by photosynthesis on the global scale and one has some idea of the trauma of the oxygen-poisoning incident when it happened.

Our capacity to pollute on a planetary scale seems rather trivial by comparison and the system does seem to be robust and capable of withstanding major perturbations. The doomsters' cliche ". . . and we'll destroy all life on earth" seems rather an exaggeration when applied to an affair such as the depletion of the ozone layer by a few percent. On the other hand, pollutants such as these which still might stimulate a minor adjustment, by Gaia, to keep homeostasis could be gravely disturbing to us as a species. To Gaia a glacial epoch is no more than a chill, for it affects no more than the top and bottom 30 percent of the latitudinal zones. Thus, 70 percent of the surface of the earth is comparatively unaffected by a glacial epoch. The other 30 percent is partially frozen anyway in the interglacials, so that one should not look on the glaciation from a planetary viewpoint as anywhere near as disastrous an event as it would be to us.

Much more serious than the "blind chance" damage from pollution is the purposeful biocidal activity of agriculture. Urban dwellers, and that includes most scientists, tend to forget that the modern farmer regards all living creatures other than his crops and livestock as weeds, pests, and vermin to be destroyed. It may be that we can escape the consequences of this sort of activity on the land surfaces. Recent measurements of the total biospheric productivity suggest that there have been no adverse effects which could be attributed to the present level of farming. I wonder, though, if farming is extended to the marshes of the world and to the continental shelves, if we might be in peril, for it would seem that those regions are an important seat of our hypothetical regulatory system. Here is buried much of the carbon, thereby sustaining oxygen; here is made much of the methane and nitrous oxide which may control it; and here, of course, are produced other trace gases, which may be important in the mass transfer of elements throughout the world. The extension of farming to these zones could conceivably be dangerous.

In addition, marine algae are also a source of a wide range of trace gases and on a scale quite large compared with the chemical industry. Thus, methyl chloride is a natural product of these regions. If we are to be concerned by the presence of chlorocarbon pollution by industry or oxides of nitrogen from combustion, we should be concerned equally about the production of these same gases from natural biological sources. Agriculture, especially if extended to marshes or continental shelves, might alter drastically the natural production rates, one way or the other.

The consequences of our presence on the planet, therefore, seem to be largely a matter of scale. We now manipulate whole geographic regions to our short-term advantage. What would happen if we were obliged to use the whole planetary surface for food? Not long ago it would have been said that we then had won our final victory over nature and the earth would be truly ours—our spaceship with the crops and livestock as the life support system. Maybe future agronomy will be sufficiently subtle to produce a stable system of this type, but if we are right about Gaia, whoever owns the planet has the job of driving it, and on us then would fall the daunting, if not impossible, task of planetary engineering and the maintenance of the optimum environment that we now get for free.

Comments and Discussion

The provocative contribution by Dr. Lovelock resulted in an extensive discussion, the first part of which was concerned directly with Dr. Lovelock's presentation. He was asked whether the Gaia hypothesis required an "intelligence" to evaluate and regulate the biosphere–atmosphere system. Dr. Lovelock suggested that chemical "signals" in the biosphere could adequately transmit information and Gaia probably did not require a "brain" to function. He noted, however, that we do not yet know how Gaia operates.

Another participant commented that even if Gaia possessed some "theological" set of nonlinear negative feedback mechanisms which would mitigate many of man's impacts on the system, man still might be threatening the survivability of a few percent of his species by his own activities. Unlike certain natural biological systems which may regulate themselves with adjustments of a few percent, when we are dealing with humans the threats to the survivability of a few percent of the population, given the "nonlinearities" of terrorism, nuclear war, and other destructive tools which man may use when threatened, are one of the main reasons for our concern at this Conference.

Dr. Lovelock agreed and suggested that, by exaggeration, the "wilder" of the environmentalists weaken the case for concern. However, another participant noted that Dr. Lovelock did not emphasize strongly enough the fact that, with respect to the flows of energy and materials, civilization has become, or is rapidly becoming, a force comparable to natural global processes. The fact, for example, that organisms have dealt with and transformed mercury over millennia should not make us entirely complacent about our current mobilization of mercury at much larger than geological rates. Though we probably are not threatening the survivability of the entire biosphere, changes of 5 or 10 percent in the carrying capacity of the earth for human beings must be viewed as having enormous social and political consequences on a global scale (see Part 3).

A conferee pointed out that certain of man's activities, such as Brazil's removal of the tropical rain forest or the USSR's plan to divert southward rivers that now flow into the Arctic Ocean (see Part 2), can affect the global climate through their effect on the general circulation of the atmosphere and oceans. It was suggested that the conferees should address the problem of how nations deal with one another when activities of one nation are transmitted around the world through the general circulation (see Part 5).

Dr. Lovelock stressed that he was *not* suggesting that the Gaia hypothesis implies that we could let nature solve man's problems and that everything would be all right. Instead, he pointed out, if Gaia exists, our approach to environmental problems might be somewhat different from one in which we assumed nature was a passive entity with no homeostatic response changes. Gaia might provide a rationale for global studies of environmental issues.

Regarding environmental impacts by human activity, Dr. Lovelock was asked about the natural vs. manmade flow of chlorocarbons into the atmosphere. He observed that methyl chloride is the dominant chlorocarbon in the atmosphere although we do not know whether its main source is marine algae or slash-and-burn agriculture and the burning of vegetable matter. It is not commonly known, he said, that smoldering combustion converts one percent of the chlorine content of the organic fuel to methyl chloride, an astonishingly large conversion to an unlikely substance. Until we have determined the sources of methyl chloride, it is best to leave any conclusions in abeyance.

It was pointed out that it is not the abundance of chlorocarbons in the troposphere which is important, but rather whether the chlorocarbon reaches the stratosphere where it may affect the ozone layer. As for methyl chloride, it certainly does reach the stratosphere but it is fairly stable, even there. Its concentration, about one part per billion in the stratosphere, is about 10 times that of fluorocarbon-11 and -12. A significant fact is that we only learned this year of methyl chloride's existence in the stratosphere.

A participant asked Dr. Lovelock to outline the threats to the ozone layer and the consequences of those threats to human existence. Dr. Lovelock first noted how preliminary and incomplete our knowledge of those threats is. He stated that the current threat from industrial releases probably was not too serious—a theoretically calculated maximum of one percent of the ozone layer has been removed by flux of artificial chlorocarbons into the stratosphere. Since the natural annual variability of the ozone concentration in the stratosphere is about 12 percent, there is little need for concern at this time. A participant retorted that, while changes less than the natural variability may be impossible to observe, the long-term change in the mean may be important. The chairman suggested that the ozone issue be discussed later in another, more appropriate session (see Part 2).

A conferee observed that the Gaia hypothesis might have two different impacts on the general public. One impact would be to soothe people, assuring them that since the system had worked for millions of years, we are safe from our own activities. The other impact might be to suggest that, since the system is so complex and interwoven, small changes might have enormous importance.

A crucial exchange, perhaps the fundamental problem confronted by the Conference, occurred when a participant suggested that we should convey to the public the conclusion that "we don't know enough, we are

trying to learn more, let's hang on tight while we can." Immediately came the response, "No, let's hedge against the worst."

A few scientists expressed difficulty in accepting the Gaia hypothesis. One mentioned that he had seen no real evidence that, with certain exceptions, the conditions that exist in the atmosphere were caused by anything other than inorganic processes. When reminded of two apparently lifeless planets, Mars and Venus, which had virtually no free oxygen or nitrogen, he noted that this could be due to the lack of water rather than to a lack of organic processes. Regarding optimum conditions for life, he suggested that evolution selects those organisms that live best under the present conditions on earth. It was suggested further that inorganic models simply had not been developed sufficiently to account for all processes. Dr. Lovelock responded that Gaia was a hypothesis "put up to be shot at." He opined that there was sufficient difficulty with inorganic models to suggest that Gaia could stand as a valid theory and referred again to his table showing the contrast between the organic and inorganic equilibrium concentration of the key atmospheric constituents.

Another participant suggested that perhaps organic processes compete with inorganic processes and have been driven to some less-than-optimum but steady-state level where life has come into balance with inorganic processes. It is perhaps only by chance that the oxygen content of the atmosphere is 21 percent, rather than because that is the optimum value for life. Thus, organic processes may not themselves maintain a 21 percent oxygen level as an optimum condition.

Finally, the ability of the global organism to adapt to changes produced by man on short-time scales was seriously questioned, even if the Gaia hypothesis proved correct. Is it crucial that we understand completely what effects our activities might have on the biosphere, or is it of greater importance to know whether nature can adapt to those changes on time scales short compared to evolutionary scales? The point was made that while life has existed on the earth for some 3 billion years, no one subsystem has survived that long. Thus, it is not obvious that man's survival as a species is insured by Gaia.

APPENDIX III

Some Thoughts on Control of Aerospace

RAYMOND SLEEPER
Summarized by J. Dana Thompson

In 1952, in the Air-War College at Maxwell Field near Montgomery, Alabama, a study of "air control" was begun. It was a British concept which stated that aircraft, through the control of the air, could affect the behavior of people on the earth. The study drew on research just completed at the Harvard Research Center under Dr. Kluckhohn, a social anthropologist of considerable renown, which had developed a "working" model of the Soviet air control system.

The main hypothesis of "Project Control," as it came to be called, was that the USSR, a closed social system, depended upon, among other things, a security system the West called the "Iron Curtain" as its primary means of information control. If the Iron Curtain could be penetrated and eroded effectively, it was believed there might be a more open society in the USSR.

A major, high-flying reconnaissance offensive over the USSR was postulated. The aircraft to be used initially was the Royal Air Force's Canberra, later the B-57, later still the U-2, and, finally, satellites. When briefed on this proposal the President liked and approved it, and titled it "Open Skies." The goal of the operation, as originally formulated, was sociopolitical—to erode the Iron Curtain and to open up the political system in the USSR. (As it developed, the goal became to acquire highly classified secret intelligence.) Initially, the goal was to fly so high that Soviet airspace could be penetrated with impunity; one means of gaining control of the airspace.

It is an article of faith in all air forces that in conflict their main task is to gain and maintain control of the air. The air force that flies with impunity, controls the air. "Open Skies" set out to establish a de facto "right to fly" policy anywhere over the globe *above* the sensible atmosphere, i.e., above the weather. In effect, it was a unilateral declaration of open skies. Unfortunately, it did not turn out that way. The Department of State and the Department of Defense both objected, on the grounds that our intelligence must remain highly classified. Nevertheless, the "Open Skies" project was launched and eventually both the U-2 and satellites were used. The air incursions into Soviet airspace produced very vigorous air defense, missile, satellite, and, recently, civil defense programs in that country. Had we forced the political issue of freedom of airspace then, I believe that today there would be, in effect, a global structure of controlled airspace that would be extremely useful in our present deliberations on environmental control.

When the reconnaissance satellites began to fly, it was thought for a while that the Soviets were going to shoot them down, but they did not. When the first manned space launch was made by Yuri Gagarin, he did intrude into sovereign U.S. airspace—at an altitude above 50 miles. Both nations, of course, now have satellites flying globally. These satellites have the capacity not only to photograph weather, clouds, and other features of environment but they have the potential for monitoring the environment rather thoroughly, if we care to instrument them, establish the ground monitoring and control stations, and conduct the analyses. There is much more to be done with satellites than has been done to date—over our soil as well as theirs.

Another facet of control of the airspace is air transportation. As air transportation grows, it has a tremendous impact on the structure of man's social systems. Today, there are two serious constraints to the continued growth of world air transportation—the cost of fuels and the cost of airport access. The 15 percent per year growth rate of air transportation in this nation in recent years is leveling off due to these constraints, both of which can be solved technologically. New engines, new fuels, new strong and light, fibrous plastics for aircraft construction— all can be used to battle higher fuel costs. The airport access problem requires an institutional solution.

Assume, for example, that freight is shipped from one urban area to another. About one-third of the shipping costs may be required to actually fly the freight, while the remainder is used for ground transportation expenses. Thus about two-thirds of the total cost is attributable to airport access. Why? Because we build our airports 5 to 50 miles from our metropolitan areas and then build *around* the airport to make it difficult to reach the cities.

At the University of Tennessee Space Institute (UTSI) we have been working on a plan for a multimodal, integrated airport transportation center which becomes an "air city." The city is designed so that all points are within a 30-minute drive of the airport. It is a huge airport, 80 square miles, with four parallel runways, clear approaches, noiseproof buildings where necessary, and in-airport industrial, commercial, and housing complexes. It is in effect an "air harbor," a facility that does not exist anywhere in the world today. When you go to this airport, you go to the city. It is quite different from Dulles International Airport, 25 miles from Washington, D.C., or the Dallas–Fort Worth Airport, 40 miles from Dallas.

Air cities are being built *now*. They are not designed, however, they simply grow, like Topsy. Chicago's O'Hare Airport is a good example.

Look at a photograph of O'Hare taken from 40,000 feet. Draw a circle 25-miles in radius around O'Hare, then draw a similar circle around Chicago's central business district. Compare the two. At present growth rates, O'Hare "city" will be larger than Chicago by 1985. But it was not designed as a city, it was designed as an airport for Chicago.

The point is, the technology of the wide-bodied jets, the SST, the new engines, new aircraft and materials, and new aerodynamic principles now make the million-pound aircraft feasible and economically desirable. But we cannot capitalize on this new technology until we design our social systems to be compatible with it.

This brings us to the new aircraft. First, the SST. I believe the possible social benefits of the SST are worth its whole investment. The United States invested nearly one billion dollars on an SST and then cancelled it. The issue was complex, but one thing that forced its cancellation was the environmental factor. Some claimed it would destroy the ozone layer and cause an increase in skin cancer; so Congress stopped the SST program. But Britain, France, and the USSR all have built supersonic transports—and our own subsequent Climatic Impact Assessment Program has stated that these SST's probably will not harm the atmosphere. Many people opposed the SST, mostly on the basis of fear and suspicion. It now appears that most of that fear may have been unfounded. Further, the United States had a real technological lead on the world in SST development. Today, Britain, France, and the USSR have built and are flying SST's, and we are going to have to decide whether they will be allowed to operate regularly from U.S. airports.

The SST is twice as productive as the Boeing 747, but, more important, the SST has the potential of bringing the peoples of the world within 2 to 3 hours of each other. If we cancel a new technological development because we fear that it will harm the environment—while the rest of the world goes on—we suffer economically. We must be sure that the technology is completely incompatible with the environment or we must alter the technology to make it compatible with the environment. It does not make sense to drop out of the competition because of fear.

The same general point is true for nuclear electric power systems. The parallels with the SST controversy are rather clear and we need not go into detail. However, this leads logically into a discussion of nuclear weapons and nuclear war, a subject little discussed at this Conference, but a real threat to our social system and our airspace.

The detonation of nuclear weapons already has placed some 5 tons of plutonium into the environment. To emphasize the reality of nuclear weapons, consider that the USSR now possesses about 3,700 nuclear bombs, compared to the U.S. stockpile of 2,100. In addition, the United States has some 7,000 tactical nuclear weapons in Europe against some

3,500 such nuclear weapons of the USSR, not to mention the increasing nuclear capability of Britain, France, China, and now India. But that is not all. The Soviet Union is testing four new intercontinental missiles, the SS16, SS17, SS18, and SS19. New evidence indicates that at least four *additional* weapons systems are about to be tested. The Soviets apparently are committed to augmenting their nuclear defenses.

If we have even a very limited nuclear war, our airspace and everything in it will suffer, to say the very least. Of great concern is evidence that the Soviets are strengthening their missile facilities and conducting a vigorous civil defense program. The Soviets preach that in a nuclear war they would "win." We must consider very seriously the nuclear war–nuclear weapons situation.

Where does all this lead us? I suggest that the overall direction be one of *balance*. It is rather difficult to define a global environmental policy of balance when the leading technological nation has not yet established a balanced energy policy. But it would seem that one thing we might formulate here is a step toward an initial *balanced policy statement* about a global environment.

APPENDIX IV

International Structures for Atmospheric Problems

JACK E. RICHARDSON
Summarized by J. Dana Thompson

The International Civil Aviation Organization has been a fairly effective body in regulating the technical aspects of aviation. But it has broken down when dealing with the "hip-pocket" arrangements of particular countries who wish to maintain air transport industries. The best results in trying to obtain efficient use of airlines have been achieved through regional organizations. Agreements in Europe, for example, have been made on nonscheduled flights to rationalize or standardize equipment. Yet we are a long way from achieving the best economic use of our technology. The average load factor in international air transport is well under 50 percent of total carrying capacity. The rational, scientific solution to this wasteful situation would be to *internationalize* the major trunk-route air systems, thereby increasing aircraft use and forcing aircraft manufacturers to produce exactly the kind of aircraft needed. Certainly that is a desirable goal—but the chances of achieving it are negligible. The reason is rather simple—national interests. We all love to have our own flag-carrying aircraft and, moreover, by doing so we provide a training ground for air crews, enhancing our defense potential. Our rational, scientific objectives are thus frustrated by national interests.

It is clear that when dealing with environmental issues, we must turn our attention from national sovereignty to common problems of the environment, irrespective of national boundaries. Somehow we must evolve a mechanism to work in that direction. One of the difficult problems in reaching a multilateral convention on the environment, which probably would require countries to take quite stringent action, is that many countries have not even begun legislating against their own domestic environmental problems. The United States has, through the National Environmental Policy Act, at least recognized that it is a polluter. But, unfortunately, there are not many repetitions of that kind of legislation in other countries. Many countries have achieved nothing in terms of protecting their own domestic environment. How then can we hope to achieve, at the international level, a convention that would impose quite severe economic and other controls on the participating countries?

Imagine, for example, an international convention which required

Japan to cut down its contribution to global atmospheric CO_2 content by decreasing industrial production. That could have severe effects on the national welfare of Japan. It is unrealistic to expect countries like Japan to sacrifice their current economic welfare for the sake of long-term environmental results of which other countries are the beneficiaries, even though we think they should. Such very real problems are inevitable in a multilateral convention on environment.

Perhaps a better initial approach would be to set out in a document the nature of the possible impacts of civilization on the environment and possible remedies and safeguards—in short, an impact study which presents the spectrum of scientific opinion as to where the dangers to mankind lie. This must be a first step before meaningful results at an international level can be obtained. In contrast to a conference on disarmament, where all parties recognize why one needs to reduce armaments, the case for environmental concern on an international level has to be made. Nations first have to be encouraged to take their own domestic action on environmental problems. The need to identify the problems and risks to humanity due to pollution of the atmosphere must be publicized.

Some progress has been made on a much more limited scale. The bilateral draft U.S.–USSR Treaty on Environmental Modification, made public recently, is one step forward. Each state agrees not to engage in environmental modification for military purposes to the harm of other countries. While these bilateral agreements can achieve a great deal, they usually result from polarized attitudes rather than from any essential conviction of the community of nations that what they are doing is for the good of mankind, however it is dressed up. It is a start, but it is doubtful whether global environmental control can be accomplished by such bilateral methods.

Assuming that we can identify the environmental problems, it may be useful to promote a regional organization of nations to consider environmental issues. While the exact grouping is not obvious, it is quite clear that the countries of South America, for example, have a different idea of the dangers of pollution and environment than the countries of Europe, or the South Pacific, or North America. A better starting point, therefore, appears to be arrangements of regional conferences at the governmental level.

It is also possible to begin at "ground level." For example, COSPAR (Committee on Space Research) came into existence to foster the international cooperation in space research that began during the International Geophysical Year. COSPAR consists of representatives of a good many national scientific organizations, set up on a nongovernmental basis with an executive bureau which could translate the policies of COSPAR into practice. That bureau, although dominated by the United

States and the USSR, is a kind of model that ought to be looked at in relation to the things we are trying to achieve—for example, disseminating advice to governments for some possible form of subsequent executive action. Of course, that introduces the governmental element of control over the policy activities of the body.

An alternative suggestion is to create a nongovernmental or mixed body in the form of an institute that can examine the problems we have been discussing in greater depth and greater breadth, with emphasis on greater breadth. Membership in such an international institution would include scientists, economists, and lawyers. This body would be set up on a permanent basis to produce results which could influence governments and lead to regional and/or global conferences on the environment.

An example of such an institute, the David Davies Memorial Institute in England, embarked in the 1960s upon a program of drafting a code of rules for the exploration and use of outer space. The Institute, an assembly of primarily British lawyers, scientists, and engineers, produced a number of suggestions and provisions which subsequently appeared in quite similar form in the Outer Space Treaty of 1967. That kind of approach is very useful—but results are not achieved overnight. Admittedly, it is not a very "high-voltage" approach, but it is worth considering.

Comments and Discussion

Following Professor Richardson's presentation a number of Conference participants made useful comments. It was noted that the institute idea had been used in the creation of a research organization to study the common problems of the more advanced societies. The charter for the International Institute for Applied Systems Analysis (IIASA) was drawn up 3 years ago. Its membership includes all of the eastern European countries, a number of western European countries, the United States, Canada, and Japan. It has been established at Laxenburg, Austria. It was suggested that this Institute might undertake the type of studies suggested by Professor Richardson.

The useful role of COSPAR as an advisory group was elaborated. It was noted that COSPAR organized the Consultative Group for Potentially Harmful Effects of Space Experiments, which was asked by the United Nations to study several questions related to the effects of various activities on the upper atmosphere. Useful answers to these questions were provided to governments as a result of the Group's study.

It was suggested that the Conference specify in detail what the objectives of the proposed international organization might be and what functions and activities it might carry out (see Part 5). One participant expressed concern that the WMO might be "politicized" in the process of constructing an international organization concerned with environmental issues. The WMO was thought to serve best as a consultative group on technical questions. Professor Richardson's suggestion that environmental problems first be addressed on the regional scale was questioned as being impractical from an international economic standpoint. Regions imposing environmental restrictions and sanctions against themselves and their industries probably would become uncompetitive in the international marketplace. A case in point concerns the "acid rain" problem of the Scandinavian countries. Industrial polluters of Northern Europe claim they readily would impose emission controls if all their international competitors were required to do likewise.

Finally, a participant noted how ineffective large organizations are in looking ahead, anticipating next year's problems. How can we establish a climate in which the curious investigator working in an obscure discipline is assured of support? It was suggested that the Conference address the problem of how "unfashionable" but potentially fruitful research might be funded through an international organization.

APPENDIX V

Some Comments on Technology and the Atmosphere

JOHN STROUD
Summarized by J. Dana Thompson

I am a general scientist, by my own choosing, and belong to no recognized or established profession. I am basically interested in the evolution of superhuman control systems. The objects of my interest, if you were speaking to a historian, would include "Roman Civilization," "Chinese Culture," and the "Western World." In anthropological terms, these specific interests would be characterized as "agrarian systems," with perhaps a whole string of qualifying adjectives. I call them Type III systems.

Basically, the difference lies in the paradigm used; the things assumed, the hypotheses generated, or the data collected to test them. In my case I use a form of systems engineering paradigm which I early learned to call "functional." In this paradigm a system *is* what it *does*. More precisely, it is a domain of controlled action.

To illustrate this, consider my friend Joe, who has been out of town for several months. When I next see Joe, I ask each atom in his body to give his name, rank, serial number, duty, and duty station. When I do this, if I do not know already, I may be very surprised to find that many of Joe's atoms have "gone over the hill," since residence time of a typical atom in the human body is of the order of days. The few atoms still remaining have an entirely different job, defined by distinctive quasi-stable state relations with other atoms.

Yet, I will not respond to the atoms' answers by asking incredulously "Are you Joe?" Of course he is Joe! He acts like Joe. That is the only meaning to a system named Joe there is in this analysis.

Superhuman systems are extremely interesting, especially when you consider that they have controlled the actions of so many people over the time scale of many centuries and over length scales of thousands of kilometers. They are even more intriguing when you realize that some of them have quite long lifetimes.

I stated I was interested in the evolution of superhuman control systems. This was not my original interest. I simply found that to make a stumbling jump a thousand years into the future I had to go backwards 2 million years! Thus, in my current way of looking at things, I find it convenient to characterize mankind Type I. Unless it should transpire that the "Yeti" of Tibet and the "Sasquatch" of British Columbia are really living fossils, you will not find a Type I representative alive today. Type II hunting and gathering systems are still on the globe, but are declining in numbers. Type III systems abound. Historically, we know much about a number of them because they left records of their existence. There were also Type III's that anthropologists would call "Neolithic," who never got around to writing anything down.

What about Type IV systems? We do not have a common name for those systems yet, although we talk vaguely about "modern science and technology" or "the space age."

To illustrate one possible difference between a successful Type III and a successful Type IV human consider the following: A typical Type III might acquire 47 different ways to kill, prepare, eat, and enjoy most anything on earth that is alive and big enough to see. A Type IV might have a typical diet consisting of sunshine, moondust, and asteroid chips—just as palatable and which maintains physiology just fine. I believe that most of the crises which we face today are related to this transition from one type of superhuman control system to another.

There are already a number of primitive, infantile Type IV superhuman control systems. Yet we are not aware they are Type IV's because we call them by very specific names—the Polaris Weapons System, the Apollo System, or similar titles.

Let us pretend for a moment that such Type IV systems are viewed in the perspective of a Type III system.

Consider the Polaris Weapons System. If you examine the number of man-hours per day this system controls (and very strictly), you would discover that the number corresponds to that expended in one of the smaller nations of the world on *all* activities. If you were to apply a measure such as Gross National Product to the Polaris System, you would discover that the GNP per citizen of the culture of Polaris was several times your own.

What does the Polaris Weapons System do? It is a fairly effective symbolic threat used against other nations by the United States. Fortunately, no one has died of it. Unfortunately, it has another function which we hope will never be excited. For if you call upon this system to act in its second mode—within the next hour, "without half trying," more people than live in Canada could be dead, and by the end of a year, more people than live in the United States could die as a result of the system.

Now assume we examine a Type III system from the perspective of a Type IV system. Had we designed a Type III system as we go about designing Type IV's, there would be most notable differences. It would have a much shorter than normal life span, for example. Type IV systems are conceived and developed, mature and age into obsolescence. In fact, we probably have spent more money studying the aging of missile systems than on the aging of humans. The strange thing is that Type IV systems are *expected* to die. And designers worry about their successors—what kind of descendants will they have, what will they be able to do? You never heard of a nation (Type III) planning its own funeral or worrying about its cultural children.

Type IV systems have unusual effects on human behavior. Consider the space colony system, being proposed by Gerard O'Neil and his disciples. In that system there are no absolutely right ways to do anything—the right way is the one that gets the job done. There are no things that you are supposed to be—in that world you will be what you have to be to get things done. There are no invariant requirements over time—the right way to do something is the way that works and that is only so long as a function needs to be performed. It is a much different kind of world—a Type IV.

For the moment, let us address the practical problem of how Type IV systems are financed and come into existence. Consider a proponent of a Type IV system such as the space colony. He has a particular value system—a set of rules for making decisions. However, his value system probably is not identical with that of the Type III superhuman control system in which he is embedded, the United States, for instance. How does the proponent influence the decisionmaking process of the Type III system on behalf of his Type IV conception? He must persuade the implementers of the Type III system in terms which correspond to the way they, the implementers, see the universe. He must persuade them on terms they understand, in ways consistent with their rules of decision-making. In the process it is quite impossible for the proponent to tell the truth precisely as he sees it.

A well-known example of this approach is the case of Dr. Werner von Braun and his colleagues in Germany. Before World War II, this group was very small, dedicated to rocketry, and headed for the moon. Unfortunately, their Type III system was led by someone who wanted to buy a nuclear weapon. So a weapon was sold but a trip to the moon could not be sold. Later on, with their former system destroyed, some of them, including von Braun himself, found a new Type III, the United States. Initially, the military of this system recognized how important von Braun's work was to them and supported it. Still later, when the United States began to lose face in the world as a result of Russian space successes, the United States once again found an important Type III use for von Braun and some of his colleagues. National pride and vanity are extremely important in Type III systems.

Eventually, von Braun realized his dream to reach the moon. Yet the means for getting there depended crucially on a Type III system sponsoring his Type IV system development—and it did so for reasons of self-interest, for the good of the Type III system.

A related practical problem for Type IV systems is their short life span—not because of their intrinsic problems, but because they depend on Type III systems which are notoriously fickle.

Now this Conference has a rather similar practical problem. I submit that you, too, want to do something which requires considerable positive and negative resource commitments over which you do not have control—you are not political or economic moguls. How do you convince the implementers of Type III systems that they should commit resources they control? First, you should recognize that these implementers represent the body of the Type III system—taxpayers, if you will. They may reward you with support if you can demonstrate a benefit to them and their lifestyle from your activities. You must show them a need and a benefit. So keep your public relations material in your sales brochure.

You may not have to solve the problem you appear to face today. The game may change and the solutions to the presently apparent problems may, by then, be irrelevant. More important problems may evolve for you to attack. But the practice will do you good.

In closing, I would like you to imagine yourself as "Mr. Big," an important Type III implementer who controls the system's resources. Are you sure you need what the Type IV proponents are selling? Do you suppose you could persuade all of them to go away—and take their filthy hardware with them? Sometimes I wonder about that myself.

Rapporteur's Observations

Mr. Stroud's presentation was, in the words of one participant, "strongly moving." He offered to the Conference a straightforward, practical, and rather cynical collection of guidelines, anecdotes, and personal experience concerning the general problem of how scientists can persuade their society to sponsor research on important scientific and social problems. Due to time limitations, comments and questions from participants were postponed until the end of the session.

Bibliography

BACASTOW, R., AND C. D. KEELING. 1973. Atmospheric carbon dioxide and radiocarbon in the natural carbon cycle: II. Changes from A.D. 1700 to 2070 as deduced from a geochemical model. Pages 86–135 *in* G. M. Woodwell and E. V. Pecan, eds. Carbon and the Biosphere. U.S. Atomic Energy Commission, Washington, D.C.

BRODERICK, A. J. 1976. Effects of cruise-altitude pollution. *J. Aircraft* (in press).

BROECKER, W. S. 1975. Climate change: Are we on the brink of a warming? *Science* 189:460–463.

BROECKER, W. S., AND A. KAUFMAN. 1965. Radiocarbon chronology of Lake Lahonton and Lake Bonneville II, Great Basin. *Geol. Soc. Am. Bull.* 76:537–566.

BROECKER, W. S., Y. H. LI, AND T. H. PENG. 1971. Carbon dioxide—Man's unseen artifact. Pages 287–324 *in* D. W. Hood, ed. Impingement of Man on the Oceans. Interscience Division, John Wiley & Sons, New York.

BROOKS, C. E. P. 1970. Climate Through the Ages. 2nd revised edition. Dover Publishing Inc., New York.

BROSSET, C. 1975a. Ammonium sulfate aerosols. Swedish Water and Air Pollution Res. Lab. Rep. B248, Gothenburg.

BROSSET, C. 1975b. Determination of airborne acidity with some examples of the role of acid particles in acidification. Swedish Water and Air Pollution Res. Lab. Rep. B226, Gothenburg.

BROWN, L. R. 1975. The world food prospect. *Science* 190:1054.

BRYSON, R. A. 1974. A perspective on climate change. *Science* 184:753–760.

BUDYKO, M. I. 1969. The effect of solar radiation variations on the climate of the earth. *Tellus* 21:611–619.

BUDYKO, M. I. 1971. Climate and Life. Hydrological Publishing House, Leningrad.

BUDYKO, M. I. 1972. The future climate. *Trans Am. Geophys. Union* (EOS) 53:868–874.

BUDYKO, M. I. 1974. Climate and Life (English edition, D. H. Miller, ed.) Int. Geophys. Ser. Vol. 18. Academic Press, New York and London.

BYERLY, T. C. 1975. Nitrogen compounds used in crop production. Pages 377–382 *in* S. F. Singer, ed. The Changing Global Environment. D. Reidel Publishing Co., Boston, Mass.

CHAPMAN, S. 1930. A theory of upper atmospheric ozone. *Mem. Roy. Meteorol. Soc.* 3:103–125.

COMMONER, B. 1975. Threats to the integrity of the nitrogen cycle. Pages 341–366 *in* S. F. Singer, ed. The Changing Global Environment. D. Reidel Publishing Co., Boston, Mass.

COX, R. A., A. E. J. EGGLETON, R. G. DERWENT, J. E. LOVELOCK, AND D. H. PACK. 1975. Long-range transport of photochemical ozone in northwestern Europe. *Nature* 255:118–121.

CRUTZEN, P. J. 1970. The influence of nitrogen oxides on the atmospheric ozone content. *Q. J. Roy. Meteorol. Soc.* 96: 320–325.

CRUTZEN, P. J. 1974a. Estimates of future possible ozone reduction from continued use of fluoro-chloro-methanes (CF_2CI_2, $CFCI_3$). *Geophys. Res. Letters* 1:205–208.

CRUTZEN, P. J. 1974b. Estimates of possible variations in total ozone due to natural causes and human activities. *Ambio* 3:201–210.

CRUTZEN, P. J. 1976. A two-dimensional photo-chemical model of the atmosphere below 55 km; estimates of natural and man-made ozone perturbations due to NO_x. Proc. 4th. Conf. on CIAP. Rep. No. DOT-TSC-OST-75-38. U.S. Dep. Transportation, Washington, D.C.

CRUTZEN, P. J., I. S. A. Isaksen, and G. C. Reid. 1975. Solar proton events: Stratospheric sources of nitric oxide. *Science* 189:457–459.

DANSGAARD, W., S. J. JOHNSEN, H. B. CLAUSEN, AND C. C. LANGWAY, JR. 1971. Climatic record revealed by the Camp Century ice core. Pages 37–56 *in* K. K. Turekian, ed. The Late Cenozoic Glacial Age. Yale Univ. Press, New Haven, Conn.

DANSGAARD, W., S. J. JOHNSEN, H. B. CLAUSEN, AND N. GUNDESTRUP. 1973. Stable Isotope Glaciology. Københadn, C. A. Reitzels Forlag.

DETTWEILER, J., S. A. CHANGNON, JR. 1976. Possible urban effects on maximum daily rainfall rates at Paris, St. Louis and Chicago. *J. Appl. Meteorol.* 15:517–519.

EHRLICH, P., AND A. EHRLICH. 1970. Population Resources Environment: Issues in Human Ecology. W. H. Freeman, San Francisco, Calif.

EKDAHL, C. A., AND C. D. KEELING. 1973. Atmospheric carbon dioxide and radiocarbon in the natural carbon cycle: I. Quantitative deductions from records at Mauna Loa Observatory and at the South Pole. *In* G. M. Woodwell and E. V. Pecan, eds. Carbon and the Biosphere. Brookhaven National Laboratory, U.S. Atomic Energy Commission. CONF–720510.

GROBECKER, A. J., et al. 1974. Report of findings on the effects of stratospheric pollution by aircraft. Rep. No. DOT–TST–75–50. U.S. Dept. Transportation, Washington, D.C.

GUSTAVSON, M. R. 1975. Dimension of world energy. Rep. M75–4, Mitre Corporation, McLean, Va.

HOLDREN, J. P., AND P. EHRLICH. 1974. Human populations and the global environment. *Am. Sci.* 62:288.

IMOS. 1976. Fluorocarbons and the environment. Rep. Federal Task Force on Inadvertent Modification of the Stratosphere. Council on Environmental Quality, Council for Sci. and Technol., Washington, D.C.

JOHNSTON, H. S. 1971. Reduction of stratospheric ozone by nitrogen oxide catalysts for supersonic transport exhaust. *Science* 173:517–522.

JOHNSTON, H. S. 1975. Global ozone balance in the natural stratosphere. *Rev. Geophys. Space Phys.* 13:647–649.

KAHN, H., W. BROWN, AND L. MARTEL. 1976. The Next 2000 Years: A Scenario for America and the World. Wm. Morrow and Co., New York.

KEELING, C. D., A. E. BAINBRIDGE, C. A. EKDAHL, P. GUENTHER, AND J. F. S. CHIN. 1973. Atmospheric carbon dioxide variation at Mauna Loa Observatory. *Tellus* 25:145–154.

KELLOGG, W. W. 1974. Long range influences of mankind on the climate. Presented at the World Conference "Toward a Plan of Action for Mankind." Inst. de la Vie, Paris, France.

KELLOGG, W. W. 1974. Mankind as a factor in climate change. *In* E. W. Erickson and L. Waverman, eds. The Energy Question. Univ. Toronto Press.

KELLOGG, W. W. 1975. Climate change and the influence of man's activities on the global environment. Pages 13–23 *in* S. F. Singer and D. Reidel, eds. The Changing Global Environment. Dordrecht, Holland.

KELLOGG, W. W., AND S. H. SCHNEIDER. 1974. Climate stabilization: For better or worse. *Science* 186:1163–1172.

KELLOGG, W. W., J. A. COAKLEY, AND G. W. GRAMS. 1975. Effect of anthropogenic aerosols on the global climate. Pages 323–330 *in* Proc. WMO/IAMAP Symp. on Long-Term Climate Fluctuations, Norwich, United Kingdom. WMO Doc. 421. Geneva, Switzerland.

LANDSBERG, H. 1974. Inadvertent atmospheric modification through urbanization. Pages 726–763 *in* W. N. Hess, ed. Weather and Climate Modification. John Wiley & Sons, New York.

LANDSBERG, H. E. 1975. The city and the weather. *The Sciences* October:11–14.

LEITH, C. E. 1971. Atmospheric predictability and two-dimensional turbulence. *J. Atmos. Sci.* 28:145–161.

LORENZ, E. N. 1970. Climate change as a mathematical problem. *J. Appl. Meteorol.* 9:325–329.

LOVELOCK, J. E. 1974. Atmospheric halocarbons and stratospheric ozone. *Nature* 252:292–294.

LOVELOCK, J. E., AND L MARGULIS. 1974. Atmospheric homeostasis by and for the biosphere: The Gaia hypothesis. *Tellus* 26:2–10.

MACHTA, L. 1973. U.S. Atomic Energy Commission Report, Brookhaven Laboratory (AEC Conf.-720510).

MACHTA, L., AND K. TELEGADAS. 1974. Inadvertent large-scale weather modification. Pages 687–726 *in* W. N. Hess, ed. Weather and Climate Modification. John Wiley & Sons, New York.

MANABE, S., AND R. T. WETHERALD. 1967. Thermal equilibrium of the atmosphere with a given distribution of relative humidity. *J. Atmos. Sci.* 24:241–259.

MANABE, S., AND R. T. WETHERALD. 1975. The effects of doubling the CO_2 concentration on the climate of a general circulation model. *J. Atmos. Sci.* 32:3–15.

MARSHALL, J. R. 1972. Precipitation patterns of the U.S. and sunspots. Thesis. Univ. Kansas.

MCDONALD, J. 1971. Relationship of skin cancer incidence to thickness of ozone layer. Congressional Record 117 (19 March):3493.

MCELROY, M. B. 1975. Chemical processes in the solar system. *In* D. Herschbach, ed. M.T.P. International Review of Science.

MCELROY, M. B., J. W. ELKINS, S. C. WOLFSY, AND Y. L. YUNG. 1976. Sources and sinks for atmospheric N_2O. *Rev. Geophys. Space Phys.* 14:143–150.

MCQUIGG, J. D., L. M. THOMPSON, S. LE DUC, ET AL. 1973. The influence of weather and climate on United States grain yields: Bumper crops or droughts. National Oceanographic and Atmospheric Administration (NOAA), Washington, D.C.

MEADOWS, D., ET. AL. 1972. The Limits of Growth. Potomac Associates, Washington, D.C.

MITCHELL, J. M. 1971. The effect of atmospheric aerosols on climate with special reference to temperature near the earth's surface. *J. Appl. Meteorol.* 11:651–657.

MITCHELL, J. M., JR. 1971. *In* W. H. Mathews, et al., eds. Man's Impact on Climate. MIT Press, Cambridge, Mass.

MOLINA, M. A., AND S. H. ROWLAND. 1974. Stratospheric sink for chlorofluoromethanes: Chlorine atom-catalized destruction of ozone. *Nature* 249:810–812.

MÖLLER, F. 1963. On the influence of changes in CO_2 concentration in air on the radiative balance of the earth's surface and on the climate. *J. Geophys. Res.* 68:3877–3886.

NAMIAS, J. 1972. Large-scale and long-term fluctuations in some atmospheric and oceanic variables. *In* D. Dyrssen and D. Jagner, eds. Nobel Symposium 20 (The Changing Chemistry of the Oceans). John Wiley & Sons, New York.

National Academy of Sciences. 1975. The Environmental Impact of Stratospheric Flight. NAS, Washington, D.C.

PRABHAKARA, C. 1963. Effects on non-photochemical processes on the meridional distribution and total amount of ozone in the atmosphere. *Mon. Weather Rev.* 91:411–431.

RAMAGE, C. S. 1976. Prognosis for weather forecasting. *Bull. Am. Meteorol. Soc.* 57:4–10.

RAMANATHAN, V. 1975. Greenhouse effect due to chlorofluorocarbons: Climatic implications. *Science* 190:50–52.

REVELLE, R. 1974. Food and population. *Sci. Am.* 231 (3) :160.

ROBERTS, W. O. 1974. Towards a global food policy. Draft article for *Die Belt.* 22 September 1974. Aspen Inst. for Humanistic Studies, Aspen, Colo.

ROBERTS, W. O. 1975. Key issues of interest to DOD in weather and climate change. Memo. to S. Buchsbaum, Chairman, Defense Science Board, 3 March.

ROBERTSON, D. F. 1972. Solar ultraviolet radiation in relation to human sunburn and skin cancer. Ph.D. Dissertation, Univ. Queensland, Brisbane, Australia.

ROBINSON. G. D. 1975. Weather and climate forecasting as problems in physics and statistics. Center for Environment and Man. Rep. No. 4141–531 (Final Rep. on NSF Grant GA–28417) , Hartford, Conn.

ROTTY, R. M., AND J. M. MITCHELL, JR. 1974. Man's energy and world climate. Inst. for Energy Annu. Rep., Oak Ridge Assoc. Univs.

RUDERMAN, M. A. 1974. Possible consequences of nearby supernova explosions for atmospheric ozone and terrestrial life. *Science* 184:1074–1081.

RUDERMAN. M., AND J. CHAMBERLAIN. 1975. Origin of the sunspot modulation of ozone: Its implication for stratospheric NO injection. *Planet. Space Sci.* 23:247–268.

SCHIPPER, L. 1976. Raising the productivity of energy utilization. *Annu. Rev. Energy.* 1:455–517.

SCHNEIDER, S. H. 1972. Cloudiness as a global climatic feedback mechanism: The effects on the radiation balance and surface temperature of variations in cloudiness. *J. Atmos. Sci.* 29:1413–1422.

SCHNEIDER, S. H. 1975. On the carbon dioxide-climate confusion. *J. Atmos. Sci.* 32:2060–2066.

SCHNEIDER, S. H. (with L. E. Mesirow) . 1976. The Genesis Strategy: Climate and Global Survival. Plenum Press, New York and London.

SCHNEIDER, S. H., AND R. D. DENNETT. 1975. Climatic barriers to long-term energy growth. *Ambio* 4:65–74.

SCHNEIDER, S. H., AND T. GAL-CHEN. 1973. Numerical experiments in climate stability. *J. Geophys. Res.* 78:6182–6194.

SCHNEIDER, S. H., AND W. W. KELLOGG. 1973. The chemical basis for climate change. *In* S. I. Resool, ed. Chemistry of the Lower Atmosphere. Plenum Press, New York.

SELLERS, W. D. 1969. A global climatic model based on the energy balance of the earth-atmosphere system. *J. Appl. Meteorol.* 8:392–400.

SELLERS, W. D. 1973. A new global climatic model. *J. Appl. Meteorol.* 12:241–254.

SMAGORINSKY, J. 1974. Global atmospheric modeling and numerical simulation of climate. Pages 633–686 *in* W. N. Hess, ed. Weather and Climate Modification. John Wiley & Sons, New York.

SMIC Report 1971. Inadvertent Climatic Modification: Study of Man's Impact on Climate. MIT Press, Cambridge, Mass.

Steinhart, C. E., and J. S. Steinhart. 1974. Energy: Sources, Use and Role in Human Affairs. Duxbury Press, Belmont, Calif.

Sundararaman, N., et al. 1975. Solar ultraviolet radiation received at the earth's surface under clear and cloudless conditions. Rep. No. DOT–TST–75–101. U.S. Dep. Transportation, Washington, D.C.

Thompson, L. M. 1969. Weather and technology in the production of corn in the U.S. corn belt. *Agronomy J.* 61:453–456.

Thompson, P. D. 1961. Numerical Weather Analysis and Predictions. Macmillan Co., New York.

United Nations. 1976. Pages 2–3 *in* World Energy Supplies 1950–1974. Series J. No. 19 Stat. Pap. Dep. Econ. Soc. Affairs, United National, New York.

Weakly, H. E. 1962. History of drought in Nebraska. *J. Soil Water Conserv.* 17:271–275.

Weare, B. C., and F. M. Snell. 1974. A diffuse thin cloud structure as a feedback mechanism in global climate modeling. *J. Atmos. Sci.* 31:1725–1734.

Weinberg, A. M., and R. P. Hammond. 1972. Global effects of increased use of energy. *Bull. At. Sci.* 28 (3) :5–8, 43–44.

Whittaker, R. H., and G. E. Likens. 1973. Carbon in the biota. Pages 281–302 *in* G. M. Woodwell and E. V. Pecan, eds. Carbon and the Biosphere. U.S. Atomic Energy Commission, Washington, D.C.

Wilcox, H. A. 1975. Hothouse Earth. Praeger Publishers, New York.

Wofsy, S. C., M. B. McElroy, and Y. L. Yung. 1975. The chemistry of atmospheric bromine. Geophys. Res. Letters. 2:215–218.

World Meteorological Organization. 1975. Report of the Commission for Atmospheric Sciences. Working Group on Stratospheric and Mesospheric Problems. 8–11 September 1975. WMO, Geneva, Switzerland.

Glossary

Acid rain	Rain containing significant amounts of sulfate as a result of burning sulfur-bearing coal and oil.
Aerosols	A system consisting of colloidal particles dispersed in a gas. As an adjective, the term is applied to a method of packaging in which liquified gas, having a pressure greater than the atmosphere at ordinary temperature, sprays a liquid.
Albedo	The ratio of light reflected by a planet or satellite to that received by it, or reflectivity.
Almost-intransitivity	The property of a complex and nonlinear system whereby it remains in one model of behavior for a time and then, without any external change imposed on it, abruptly shifts to another mode for a time.
Anaerobic	Organisms or tissues that live in the absence of oxygen.
Anthropogenic	Generated by mankind.
Atmospheric conductivity	Ability of the atmosphere to conduct electricity, which depends on the existence of ions in the air.
Bicarbonate ions	Carbonate radical with one hydrogen ion attached to it.
Biosphere	The aggregate of all the regions of the earth's crust, waters, and atmosphere that can support living organisms.
Biota	Animal and plant life.
Boundary conditions	In a mathematical model, the specified inputs that are not generated by the model itself.
Catalyst	Chemical change by the addition of a substance (the catalyst) that is not permanently affected by the reaction.
Climatic "precipice"	A hypothetical threshold of climatic change beyond which irreversible or dramatic changes would occur.
Condensation nuclei	Liquid or solid particles upon which condensation of water vapor begins in the atmosphere.
Conductivity	The measure of a system's conductance of heat, electricity, or sound.
"Coriolis effect"	The apparent deflection of a body in motion with respect to the earth, as seen by an observer on the earth. In fact, it is caused by the rotation of the earth.

Cryosphere	All the snow or ice on the ground or floating in the oceans.
COSPAR	The international Committee on Space Research established by the International Council of Scientific Unions.
Cumulus clouds	Clouds caused by upward motion of moist air, usually white with well-defined structures.
Deuterium	Heavy hydrogen.
"Direct-cell heat engine"	Simple circulation in the atmosphere caused by heating one part and cooling another.
DDT	Dichlorodiphenyltrichloroethane, a contact insecticide now banned in the United States.
DNA	Deoxyribonucleic acid, nucleic acids that contain deoxyribose. They function in the transference of genetic characteristics and in the synthesis of protein.
Dynamic equilibrium	Condition of a moving fluid (atmosphere or ocean) in which there is a balance of forces and a steady state.
Feedback loops	A sequence of interactions (diagrammed as a loop) in which the final interaction influences the original one. (See "ice–albedo–temperature feedback" as an example.)
Flammability	The ability to support combustion.
Fluid dynamic system	Mathematical description of a moving fluid.
Flux	Flow (of radiation or energy) through a unit area per unit time.
Fossil energy	Energy produced by fossil fuels, i.e., those dug out of the earth.
Gaia	The ancient Greek goddess of the earth.
GCM's	General circulation models of the atmosphere.
Greenhouse effect	Warming of the lower atmosphere by an atmospheric trace gas that is transparent to solar radiation but absorbs infrared radiation from the surface that would otherwise escape to space. (Analogy to a greenhouse is poor.)
Half-life	The time required for one-half of the atoms of a given amount of a substance to disintegrate; can also be applied to aerosols that are removed gradually from the atmosphere.
Homeostasis	Maintenance of the steady state when all things in nature are in balance with each other.

Hundred-year flood	A flood of a magnitude that will occur on the average every 100 years.
Hydrologic cycle	The basis of hydrology. The "cycle" by which all terrestrial waters are derived from the oceans through evaporation and subsequent precipitation as rain, snow, hail, or sleet.
Ice–albedo–temperature feedback	A theoretical concept of a feedback loop in which the interacting elements are the area of polar ice and snow, the albedo of the polar region (depends on area of ice and snow), absorption of solar radiation (depends on albedo), temperature (depends on absorption), area of ice and snow (depends on temperature), and so forth.
Ice packs	A large area of floating ice formed over many years; pieces of ice driven together by the elements.
Interglacial period	Period occurring between times of major glacial action.
Ionizing radiation	Radiation of sufficiently short wavelength (ultraviolet, X-rays, or gamma rays) that it can ionize the atoms and molecules through which it passes.
Latent heat transport	Transport of heat by the atmosphere from equator to midlatitudes in the form of water vapor; the latent heat of evaporation is released when the water vapor condenses and falls as rain or snow.
"Little Ice Age"	The cold period that lasted in Europe, North America, and Asia from A.D. 1500 or 1600 to about A.D. 1850. (See Figs. 9 and 10).
Mesosphere	The region between the stratosphere and the thermosphere, about 50 to 80 km above the surface of the earth.
Millibar (mb)	Unit of atmospheric pressure (10^3 dynes per cm^2). Sea-level pressure is about 1015 mb.
Negative feedback	A feedback loop that causes a reduction or damping of the response of the system in which it is incorporated.
Net primary production (NPP)	See Table 4, p. 44.
NSF	National Science Foundation.
Oscillate	Vibrate
Ozone	Molecule made up of three oxygen atoms; a trace gas found mostly in the stratosphere.
Paleoclimatology	The study of climates of the past.
Parameterization	In mathematical models, a statistical or empirical relationship specified between two variables.

PCB	Polychlorinated biphenyl.
Perturbation	A displacement from equilibrium.
pH	A measure of the acidity or alkalinity of water.
Photochemical oxidant	An oxidizing trace gas, usually ozone, formed by chemical reactions in the presence of solar ultraviolet radiation. In a polluted atmosphere this is an important component of "photochemical smog."
Photosynthesis	The synthesis of organic materials in which sunlight is the source of energy.
Pluvial period	Rainy period.
Positive feedback	A feedback loop that causes an amplification of the response of the system in which it is incorporated.
"Powerpark"	An aggregation of electric generating plants in a limited area.
Radiocarbon dating	A method of determining the age of plant or animal origin by measuring the radioactivity of their radiocarbon content.
Robertson meter	An instrument for measuring solar ultraviolet radiation, usually used at ground level.
Sensible heat transport	The transport of heat by the atmosphere or oceans in the form of internal (thermal) energy of the fluid.
"Sink"	A place where something disappears; can also be a mechanism by which a substance is removed from the atmosphere.
Solar radiation constant	The total flux of solar radiation at all wavelengths at the mean distance of the earth from the sun (about 1.94 cal cm^{-2} min^{-1} or 1360 watts m^{-2}).
Solar–thermal radiation balance	The balance between average solar radiation absorbed by the planet and the average outgoing infrared radiation at the top of the atmosphere.
Solar ultraviolet erythemal radiation	Ultraviolet radiation capable of damaging essential molecules of living cells—notably DNA.
Spectral distribution	Intensity of radiation as a function of wavelength.
SST	Supersonic transport—Jet transports capable of speeds greater than the speed of sound waves through the air, i.e., "breaking the sound barrier."
Steady state	The state of a system in which its characteristics do not change with time.

Stratopause	The boundary or transition layer between the stratosphere and mesosphere.
Stratosphere	The region of the upper atmosphere extending upward from the tropopause (8 to 15 km altitude) to the stratopause (about 50 km altitude).
Synergism	The interaction of two agents which together increase each other's effectiveness.
Temperature gradient	Change of temperature with distance.
Trace constituent, gas	A minor constituent of the atmosphere, generally all those except nitrogen and oxygen.
Tropopause	The boundary between the troposphere and the stratosphere (about 8 km in polar regions; about 15 km in tropics).
Troposphere	The inner layer of the atmosphere, between the surface and to 15 km, within which there is a steady fall of temperature with increasing altitude. Nearly all clouds form and weather conditions manifest themselves within this area.
Tsunami	A tidal wave produced by a seaquake or undersea volcanic eruption.
Urban "heat island"	A city warms the air over it, and the warm dome of warm air tends to persist, even in the presence of light winds.
Vertical convection	Process whereby warmer air rises and cooler air sinks.
WMO	World Meteorological Organization, whose headquarters are in Geneva, Switzerland.

Index